AF379245

Signals and Communication Technology

Series Editors

Emre Celebi⬤, Department of Computer Science, University of Central Arkansas, Conway, AR, USA

Jingdong Chen, Northwestern Polytechnical University, Xi'an, China

E. S. Gopi, Department of Electronics and Communication Engineering, National Institute of Technology, Tiruchirappalli, Tamil Nadu, India

Amy Neustein, Linguistic Technology Systems, Fort Lee, NJ, USA

Antonio Liotta, University of Bolzano, Bolzano, Italy

Mario Di Mauro, University of Salerno, Salerno, Italy

This series is devoted to fundamentals and applications of modern methods of signal processing and cutting-edge communication technologies. The main topics are information and signal theory, acoustical signal processing, image processing and multimedia systems, mobile and wireless communications, and computer and communication networks. Volumes in the series address researchers in academia and industrial R&D departments. The series is application-oriented. The level of presentation of each individual volume, however, depends on the subject and can range from practical to scientific.

Indexing: All books in "Signals and Communication Technology" are indexed by Scopus and zbMATH

For general information about this book series, comments or suggestions, please contact Mary James at mary.james@springer.com or Ramesh Nath Premnath at ramesh.premnath@springer.com.

Erik Cuevas · Alma Nayeli Rodriguez-Vazquez ·
Beatriz A. Rivera-Aguilar ·
Jesús A. López-Luquín · Carlos Guzmán-Rosales

Image Processing with Python

Theory, Practice, and Applications

 Springer

Erik Cuevas [ID]
Universidad de Guadalajara
Guadalajara, Jalisco, Mexico

Beatriz A. Rivera-Aguilar
Universidad de Guadalajara
Guadalajara, Jalisco, Mexico

Carlos Guzmán-Rosales
Universidad de Guadalajara
Guadalajara, Jalisco, Mexico

Alma Nayeli Rodriguez-Vazquez [ID]
Universidad de Guadalajara
Guadalajara, Jalisco, Mexico

Centro de Enseñanza Técnica Industrial
Guadalajara, Jalisco, Mexico

Jesús A. López-Luquín
Universidad de Guadalajara
Guadalajara, Jalisco, Mexico

ISSN 1860-4862   ISSN 1860-4870 (electronic)
Signals and Communication Technology
ISBN 978-3-032-13284-0   ISBN 978-3-032-13285-7 (eBook)
https://doi.org/10.1007/978-3-032-13285-7

© The Editor(s) (if applicable) and The Author(s), under exclusive license to Springer Nature
Switzerland AG 2026

This work is subject to copyright. All rights are solely and exclusively licensed by the Publisher, whether
the whole or part of the material is concerned, specifically the rights of translation, reprinting, reuse
of illustrations, recitation, broadcasting, reproduction on microfilms or in any other physical way, and
transmission or information storage and retrieval, electronic adaptation, computer software, or by similar
or dissimilar methodology now known or hereafter developed.
The use of general descriptive names, registered names, trademarks, service marks, etc. in this publication
does not imply, even in the absence of a specific statement, that such names are exempt from the relevant
protective laws and regulations and therefore free for general use.
The publisher, the authors and the editors are safe to assume that the advice and information in this book
are believed to be true and accurate at the date of publication. Neither the publisher nor the authors or
the editors give a warranty, expressed or implied, with respect to the material contained herein or for any
errors or omissions that may have been made. The publisher remains neutral with regard to jurisdictional
claims in published maps and institutional affiliations.

This Springer imprint is published by the registered company Springer Nature Switzerland AG
The registered company address is: Gewerbestrasse 11, 6330 Cham, Switzerland

If disposing of this product, please recycle the paper.

Preface

Image processing is a field in computer science and engineering that focuses on the analysis, enhancement, and manipulation of images using mathematical and computational techniques. It involves tasks such as filtering noise, adjusting brightness and contrast, segmenting objects, and extracting meaningful information from the images. This technology has a profound impact on various areas, including medical imaging, where it aids in diagnosing diseases through techniques such as MRI and CT scan analysis, security, facial recognition and surveillance systems, and entertainment, by enhancing the quality of photos and videos. Additionally, image processing is critical in industries such as autonomous vehicles for object detection and navigation, as well as in environmental monitoring to analyze satellite images for climate change studies. Its versatility makes it a cornerstone of technology across multiple domains.

The objective of this book is to provide comprehensive material tailored to students and professionals studying image processing. Designed to support both theoretical learning and practical implementation, it covers a broad spectrum of essential topics. These include foundational concepts such as pixel-level operations and geometric transformations as well as advanced techniques such as spatial filtering, image segmentation, edge detection, and color image processing. By integrating detailed explanations with practical examples, the book aims to equip readers with a deep understanding of image processing methods and their real-world applications, making it an invaluable guide for mastering the subject.

This book distinguishes itself through its comprehensive coverage of essential image processing methods and its emphasis on practical implementation using Python. The integration of extensive code examples is a cornerstone of their value, addressing a fundamental reality: even readers with strong mathematical foundations can struggle to fully understand an algorithm or method without seeing it brought to life in code. This book eliminates ambiguities and enhances clarity by bridging the gap between theory and practice, making complex concepts more accessible. As readers progress from foundational topics to advanced methodologies, a computational focus ensures an intuitive and hands-on learning experience. This dual emphasis on theory and practical application not only strengthens mathematical understanding but also equips readers with the skills to implement and apply image processing techniques effectively in real-world scenarios.

Python has gained popularity in academia and industry because of its simplicity, versatility, and extensive libraries. Python stands out as an ideal teaching tool in the field of image processing for several reasons. A wide range of specialized libraries (OpenCV, scikit-image, and PIL) simplify the implementation. An interactive and engaging learning environment supported by Jupyter Notebooks. Relevance to real-world applications, equipping students with practical skills. A less complex syntax compared with languages such as C++ or Java makes advanced topics more accessible.

Written from a teaching perspective, this book is designed to serve as a comprehensive textbook for undergraduate and postgraduate students specializing in Science, Electrical Engineering, or Computational Mathematics. Its well-structured content is ideally suited to courses such as image processing, computer vision, and image analysis. Beyond the classroom, the book offers significant value to researchers in engineering fields by providing detailed yet accessible materials to enhance their knowledge and support their projects. In addition, they serve as an indispensable resource for engineers and application developers. Recognizing the diverse needs and time constraints of professionals, the book is structured so that each chapter can be read independently, enabling quick access to specific practical information. This feature makes it particularly useful in addressing real-world industrial challenges. Whether for academic learning, research pursuits, or solving practical problems in various industries, this book offers a rich, navigable resource that bridges theory and application in the fields of image processing and computer vision.

The book is structured into ten chapters, each dedicated to exploring a set of elements that collectively form essential skills in image processing. These chapters systematically guide the reader through the fundamental and advanced concepts needed to master image manipulation techniques. Chapter 1 provides a comprehensive overview of the fundamental concepts of image processing, including its various types and key principles. It introduces the concept of pixel neighborhoods, a critical aspect in understanding how images are represented and processed digitally. The chapter also covers essential functions for image manipulation available in popular libraries such as OpenCV in Python.

Chapter 2 discusses pixel operations. Pixel operations in image processing, also known as point operations, are fundamental techniques that modify the intensity or color values of individual pixels within an image. These operations treat each pixel independently, without considering its spatial relationship to neighboring pixels, making them computationally efficient and straightforward.

Chapter 3 considers the geometric transformations in image processing. Geometric operations in image processing involve the manipulation of the spatial arrangement of pixels to change the shape, orientation, or position of an image.

Unlike pixel operations, which modify individual pixel values, geometric operations focus on the spatial relationships between pixels. Common geometric transformations include translation (shifting an image in the x or y direction), rotation (rotating an image by a specified angle), scaling (resizing the image while maintaining or altering its aspect ratio), and affine transformations (combining multiple operations like rotation, scaling, and translation in a single step). These transformations are essential for tasks such as image registration, alignment, and perspective correction, where maintaining geometric consistency is crucial. Geometric operations are typically implemented using coordinate mapping techniques, such as interpolation, to accurately reposition pixel values without significant loss of detail.

Chapter 4 focuses on histograms, a fundamental concept in image processing that represents the distribution of pixel intensity values within an image. Histograms graphically depict the frequency of each intensity level, spanning from the darkest (black) to the brightest (white) pixels in grayscale images, or across the red, green, and blue channels in color images. These plots are crucial for analyzing the overall brightness, contrast, and tonal distribution of an image, offering valuable insights into its quality and structure. Histograms play a vital role in various applications, including image enhancement, segmentation, and thresholding, where precise control over pixel intensities can significantly improve visual clarity and feature extraction. To illustrate these concepts, the chapter includes multiple Python code examples, demonstrating practical implementations of histogram operations and their impact on image analysis.

Chapter 5 focuses on spatial filtering, an important technique in image processing used to modify pixel values based on the intensities of their neighboring pixels. Unlike point operations that treat each pixel independently, spatial filtering considers the spatial context of each pixel, allowing for the enhancement or suppression of specific image features. This approach is typically implemented using small, predefined matrices known as kernels or masks, which are convolved with the image to achieve the desired effect. Common spatial filters include smoothing filters, which reduce noise and blur fine details, and edge detection filters, which emphasize boundaries and sharp intensity transitions. To provide practical insights into these techniques, the chapter includes multiple Python code examples, demonstrating how spatial filtering can be applied to real-world image analysis tasks, highlighting its impact on feature extraction and noise reduction.

Chapter 6 discusses edge detection, which is a critical method used to identify and highlight boundaries or edges within an image. These edges represent significant intensity changes, often corresponding to the outlines of objects, textures, or other important structural features. The primary goal of edge detection is to extract meaningful information regarding the shapes and structures present in an image while reducing the amount of data to be processed, making it a crucial step in computer vision and image analysis. Common edge detection methods include Sobel, Prewitt, Canny, and Laplacian operators, each designed to emphasize different aspects of edge characteristics, such as gradient strength or direction.

Chapter 7 focuses on corner detection, a crucial technique in image processing used to identify distinct points within an image where the intensity changes significantly in multiple directions, forming sharp turns or intersections. These corners typically represent regions of high local variation, making them essential features for tasks like object recognition, image matching, and 3D reconstruction. Unlike edges, which capture significant intensity changes in a single direction, corners provide unique, stable points that remain consistent under various transformations, such as scaling or rotation. One of the most widely used corner detection algorithms is the Harris detector, known for its effectiveness in accurately identifying corners with high speed and robustness. To illustrate these concepts, the chapter includes multiple Python code examples, demonstrating practical implementations of corner detection techniques and their impact on image analysis.

Chapter 8 considers the Hough Transform which is a powerful feature extraction technique in image processing used to identify geometric shapes, such as lines, circles, and ellipses, within an image. It works by transforming the image data into a parameter space, where the characteristics of the desired shape are represented as mathematical parameters. For example, a straight line can be defined by its slope and intercept or, more commonly in the Hough Transform, by its distance from the origin and its angle. This approach makes the Hough Transform highly effective at detecting shapes even in noisy or partially occluded images, as it can robustly identify patterns based on the accumulation of parameter votes. The method is widely used in applications like road lane detection, medical image analysis, and automated manufacturing inspection, where accurate shape identification is critical.

Chapter 9 focuses on image segmentation, a technique in image processing that involves partitioning an image into meaningful regions or segments that share similar characteristics, such as color, texture, or intensity. The primary goal of segmentation is to transform the representation of an image into a more meaningful and analytically useful form, making it easier to interpret or extract valuable information. This process is essential for a wide range of applications, including object detection, medical image analysis, autonomous driving, and video surveillance, where accurately isolating objects or regions is crucial for further processing. The chapter includes multiple Python code examples, demonstrating practical implementations of segmentation techniques and their impact on real-world image analysis tasks.

Finally, Chap. 10 explains morphological operations which are a set of techniques that process binary or grayscale images based on their shape or structure. These operations rely on the concept of a structuring element, a small matrix used to probe and modify the spatial arrangement of pixels in an image. Common morphological operations include dilation, which expands the boundaries of foreground objects, erosion, which shrinks them, and more complex combinations like opening and closing, which are used to remove noise, fill gaps, or separate connected components. These operations are particularly useful for tasks like object detection, shape analysis, and feature extraction, as they enhance the structural aspects of an image while preserving its essential shape characteristics. Morphological techniques are widely applied in applications like character recognition, medical imaging, and industrial quality control, where accurate shape representation is critical.

We wish to thank many people who were involved in the writing of this book. We express our gratitude to Mary James and Thiyagarajan. A who supported this book project.

Guadalajara, Mexico

Erik Cuevas
Alma Nayeli Rodriguez-Vazquez
Beatriz A. Rivera-Aguilar
Jesús A. López-Luquín
Carlos Guzmán-Rosales

Competing Interests The authors have no competing interests to declare that are relevant to the content of this manuscript.

Contents

Chapter 1
Introduction

Vision can be considered as a process that allows a person to obtain a large amount of information from the environment, which in turn helps him or her to navigate and perform tasks [1]. The fact that this process, with all its problems, is attempted to be solved automatically by computers formulates one of the areas of research and development of greater investment in recent years. Computer vision is defined as any attempt focused on the development of algorithms that try to get a machine to simulate to some degree the biological process of vision.

This chapter provides an introduction to vision and image processing and introduces several concepts that will be used throughout the book. Unlike most books in this area, this book does not make an exhaustive study of the human visual system, as the authors believe that even with the technology available, it should only be considered as a motivator, keeping the appropriate distance between it and the algorithms currently proposed.

1.1 Vision and Image Processing System

A vision and image processing system consists of several subsystems that operate on a scene with the aim of interpreting some feature of interest. Figure 1.1 shows how the image processing system is divided into a number of subsystems, namely low-level processing, mid-level processing, and high-level processing.

At the low level, there are processes that are performed on the respective images for smoothing, thresholding, denoising, edge definition, etc. The processes that take place at this stage of processing are often referred to in the image processing community as preprocessing. At the middle level, processes such as edge definition and feature extraction are defined. At the high level, semantic relationships between objects in the scene description are established. In this book, most of the algorithms discussed in the chapters fall into the low and middle levels.

© The Author(s), under exclusive license to Springer Nature Switzerland AG 2026

E. Cuevas et al., *Image Processing with Python*, Signals and Communication Technology, https://doi.org/10.1007/978-3-032-13285-7_1

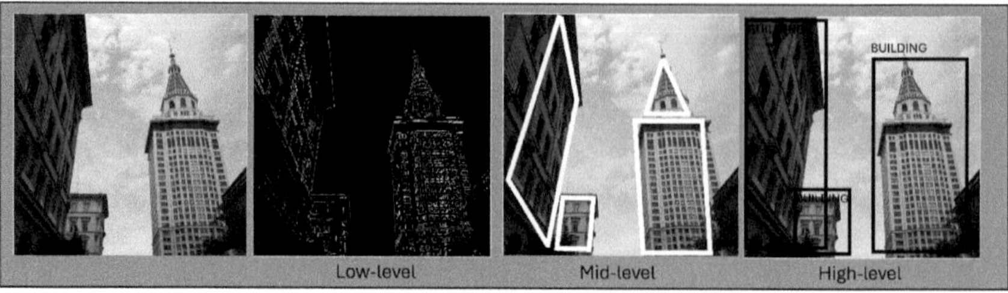

Fig. 1.1 Processes constituting the vision system

1.2 Digital Image Processing

Digital image processing can be defined as the manipulation of images by computer; the type of operations performed at the process level coincide with those discussed in the previous section [2]. An important difference between image processing and vision is the fact that the images used in the former case do not only come from the detection of the visible spectrum, which corresponds to the biological vision system, since images can come from the detection of any part of the electromagnetic spectrum. Today, there are image processing systems that work with images obtained from the detection of X-rays, gamma rays, magnetic resonance, microwaves, etc.

The raw material of machine vision and image processing is the image, which is seen as a representation of the physical world that contains important information that is captured through a sampling process, usually by electronic means.

To obtain digital images, a process of capturing, sampling, quantifying, and encoding is required. An image can be defined as a two-dimensional function that quantifies the intensity of light (the visible spectrum is the most common). An image is typically represented as $I(x, y)$, where the intensity value is obtained by indexing the x and y coordinates. The most common model of image representation is a matrix such that

$$I(x, y) = \begin{bmatrix} I(1,1) & I(2,1) & \cdots & I(N,1) \\ I(1,2) & I(2,2) & \cdots & I(N,2) \\ \vdots & \vdots & \ddots & \vdots \\ I(1,M) & I(2,M) & \cdots & I(N,M) \end{bmatrix}. \tag{1.1}$$

1.3 Basic Pixels Relationships

In this section we will establish some important relationships that exist between the pixels of an image, the idea being to describe them both in concept and nomenclature, as they will be discussed extensively in the chapters of the book.

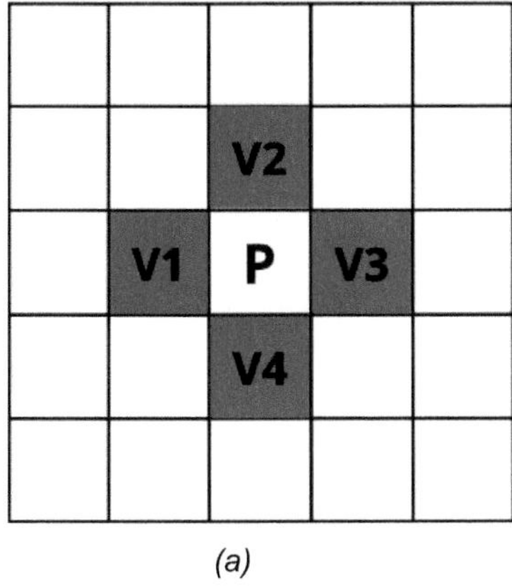

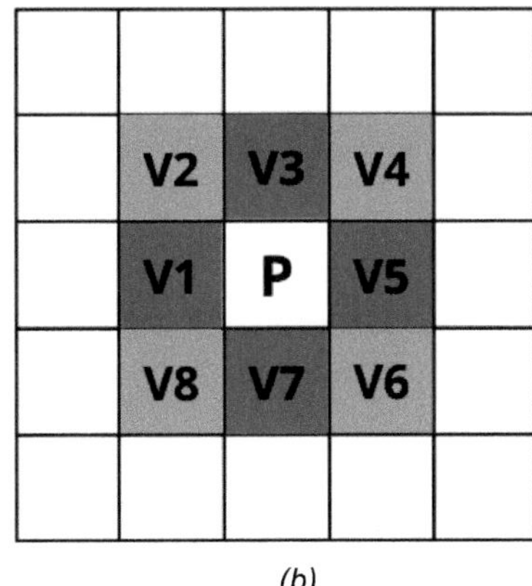

(a)   (b)

Fig. 1.2 Defining the neighborhood of a pixel P. **a** Neighborhood 4 defined over P and **b** Neighborhood 8 defined over P

1.3.1 Neighbors of a Pixel

Neighborhood is defined as the relationship a pixel has positionally to the pixels closest to it. There are two types of neighborhoods that a pixel has in the image, the 4-neighbors neighborhood and the 8-neighbors neighborhood.

The 4-neighbors neighborhood consists of the pixels (V1, V2, V3, and V4) that are above, below, to the right, and to the left of the pixel P. Figure 1.2a shows a representation of this type of neighborhood.

The 8-neighborhood is formed by the pixels (V1, V2, V3, V4, V5, V6, V7, and V8) corresponding to the 4-neighbors plus the 4 pixels diagonal to the pixel P. Figure 1.2b shows a representation of this type of neighborhood.

1.3.2 Connectivity

Connectivity between pixels is a widely used concept in the detection of regions or objects present in a given image. For this reason, connectivity is defined as a situation of adjacency and neighborhood. Under this consideration, there are two types of connectivity, connectivity-4 and connectivity-8 [3].

The concept of connectivity can be better understood by considering a binary image, i.e., an image whose pixels represent a feature rather than brightness, so their values can only be zero or one.

Given the above, two pixels and, whose values in the image are one, are said to be connected with connectivity-4 if both are in a 4-neighbor relationship. Similarly, the same pixels would be connected with connectivity-8 if both are in 8-neighbor relationship.

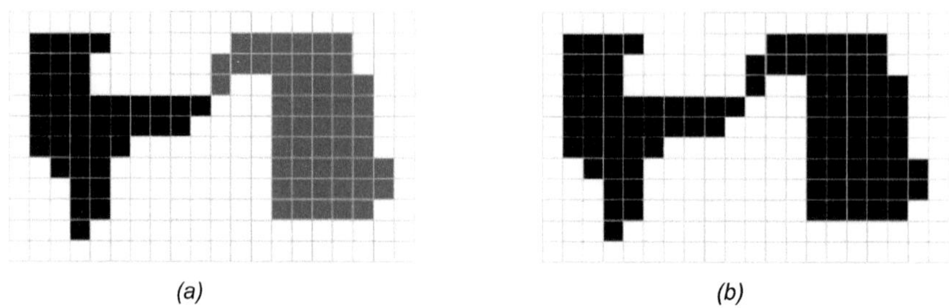

Fig. 1.3 Influence of neighborhood on pixel connectivity. **a** Image considering connectivity-4 and **b** image considering connectivity-8

To illustrate the importance of these concepts, Fig. 1.3 shows a binary image in which there are one or two objects, depending on the type of connectivity used. In the case of considering the 4-connectivity, the image will have two objects, since the point where both structures are closer, the contact pixels are not in a 4-neighbor relationship (Fig. 1.3a). If the 8-neighbor criterion is considered in the same image, the two contact pixels would be connected so that both structures would be considered as a single object (Fig. 1.3b).

1.4 Distance Measurements

The distance between two pixels is one of the most widely used measures in image processing, with applications ranging from similarity to measuring objects in the scene [4]. There are several types of measures used to find positional relationships between pixels, but the most common are the Euclidean distance, the city block distance, and the chessboard distance.

To define and characterize the distances between two points, we consider the image shown in Fig. 1.4a, which contains two pixels labeled $I_1(x_1, y_1) = p_1$ and $I_2(x_2, y_2) = p_2$.

The Euclidean distance is defined as the distance between two pixels, which is defined according to:

$$D_E(p_1, p_2) = \sqrt{(x_1 - x_2)^2 + (y_1 - y_2)^2}. \tag{1.2}$$

The distance represents the resulting vector between $I_1(x_1, y_1)$ and $I_2(x_2, y_2)$ (Fig. 1.4b).

The city block distance is the sum of the horizontal and vertical distance between the two pixels, defined as:

$$D_{CB}(p_1, p_2) = |x_1 - x_2| + |y_1 - y_2| \tag{1.3}$$

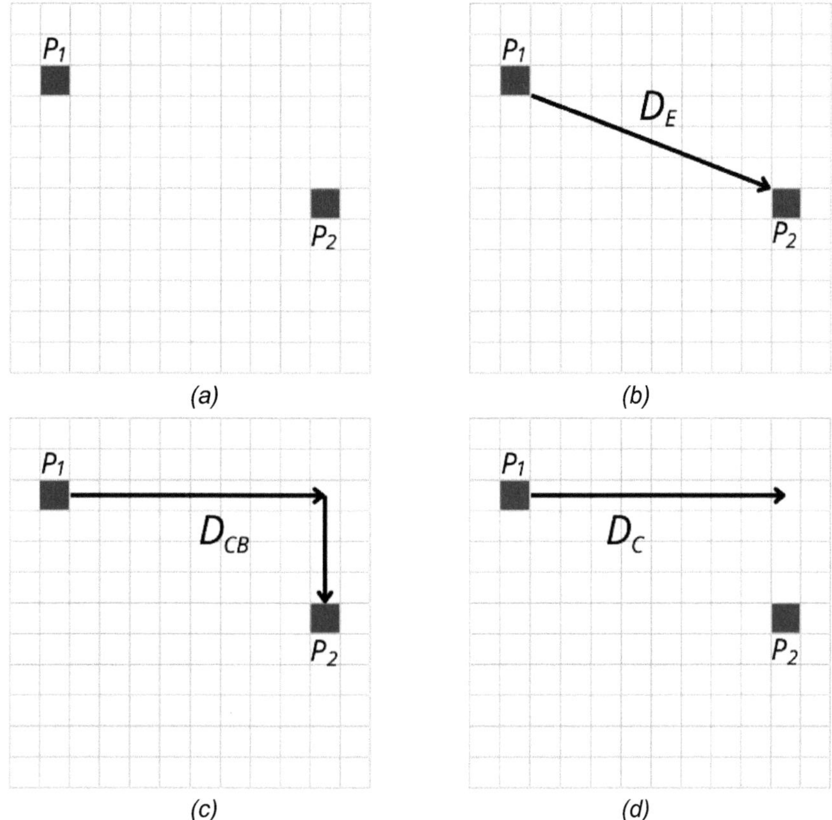

Fig. 1.4   Illustration of distance types for measuring positional relationships between pixels: **a** Pixel definition, **b** Euclidean distance D_E, **c** city block distance D_{CB}, and **d** checkerboard distance D_C

This is the distance that provides a larger value than its Euclidean and chessboard counterparts for the same positional relationship (Fig. 1.4c).

The chessboard distance is the maximum distance of horizontal and vertical displacement between two pixels. This distance is defined by:

$$D_C(p_1, p_2) = \max(|x_1 - x_2|, |y_1 - y_2|). \tag{1.4}$$

The fact that the chessboard distance only considers maximum movement in one direction makes it similar to a game of chess (Fig. 1.4d).

The decision of which type of distance to use is determined by the application and the required positional characteristics, but the Euclidean distance is the most commonly used except in those situations where characteristics such as maximum paths between pixel paths are of interest.

1.5 Read, Display, and Write Images

Until a few years ago, the image processing and computer vision community was a relatively small group with access to costly processing tools or experts in some programming languages. Today, many libraries facilitate designing vision systems, whether for manufacturing inspection applications or navigating a mobile robot. An example of such a tool is OpenCV.

OpenCV is an extensive open-source library for computer vision, machine learning, and image processing. It supports various programming languages, including Python, C++, and Java. With OpenCV, you can process images and video to recognize objects, faces, or even human handwriting. The range of available functionality expands when integrated with other libraries, such as NumPy (a highly optimized library for numerical operations). Any operation you can perform in NumPy can be seamlessly combined with OpenCV, greatly expanding your toolkit [3].

To read an image in OpenCV with Python, you use the cv2.imread() function. This function loads the image from the given file path and returns it as a NumPy array. If the image cannot be read (due to a missing file, incorrect permissions, or an unsupported/invalid format), the function will return None. The general sintaxis for this function is:

```
A = cv2.imread(path, flag)
```

where the parameter path is a string specifying the name of the image with its respective search address, and optionally the parameter flag specifies how the image should be reading from among cv2.IMREAD_COLOR, cv2.IMREAD_GRAYSCALE, and cv2.IMREAD_UNCHANGED (its default value is cv2.IMREAD_COLOR), while the variable A stores the obtained image. The image type that can be loaded with the imread function corresponds to the types described in Table 1.1.

Once the image has been loaded into a variable, it may be convenient to query its dimensions, which can be done using the shape attribute [5]. So, to get the size of the image stored in A, you would write:

```
A.shape
```

where the output will be the number of columns, rows, and channels.

Table 1.1 Image types supported by the OpenCV library

Types of files	File name extensions
TIFF	.tif or .tiff
JPEG	.jpg or .jpeg
GIF	.gif
BMP	.bmp
PNG	.png
XWD	.xwd

The cv2.imshow() function is used to display images. This function creates a window to display an image, and the window automatically adjusts to the size of the image. This way, if you wanted to display the image stored in the variable A, you would type it on the command line:

```
cv2.imshow('Title', A).
```

After processing the image, it would be convenient to save the resulting image, which is done by the cv2.imwrite() function. Its syntax is defined as:

```
cv2.imwrite('path_to_output', B)
```

where B is the image to be saved, and *"path_to_output"* is the string that defines a valid name for the image, along with an extension congruent with those defined in Table 1.1.

References

1. Gonzalez, R. C., & Woods, R. E. (2018). *Digital image processing* (4th ed.). Pearson.
2. Szeliski, R. (2022). *Computer vision: Algorithms and applications.* Springer Nature.
3. Bradski, G., & Kaehler, A. (2008). *Learning OpenCV: Computer vision with the OpenCV library.* O'Reilly Media, Inc.
4. Sonka, M., Hlavac, V., & Boyle, R. (2013). *Image processing, analysis and machine vision.* Springer.
5. Jähne, B. (2005). *Digital image processing.* Springer Berlin Heidelberg.

Chapter 2
Pixel Operations

Pixel transformations [1] are image processing techniques in which the modification of a pixel depends solely on its original value. The updated pixel $p' = I'(x, y)$ is computed directly from the initial pixel $p = I(x, y)$ located at the same coordinates. This means that the transformation does not take into account surrounding pixels or any spatial context beyond the target pixel itself.

The transformed pixel value is computed via a mapping function $f[I(x, y)]$, such that:

$$f[I(x, y)] \rightarrow I'(x, y). \tag{2.1}$$

Figure 2.1 illustrates an example of this kind of process. When, as mentioned earlier, the function $f(\cdot)$ remains constant regardless of the pixel's coordinates, meaning its outcome is uniform across all positions in the image, it is referred to as a homogeneous function. Common instances of homogeneous operations include:

- Changes in brightness and contrast across the image.
- Application of certain illumination curves.
- Inverting or complementing an image.

Some of the above examples will be discussed in detail in this chapter.

In contrast, inhomogeneous pixel operations consider the value of the pixel in question and its relative position in the image, i.e.,

$$g[I(x, y), x, y] \rightarrow I'(x, y). \tag{2.2}$$

A common image processing technique involves applying non-uniform adjustments to modify contrast or brightness based on pixel location selectively. This approach ensures that certain areas of the image undergo significant adaptation while others remain largely unaffected or receive minimal adjustments.

© The Author(s), under exclusive license to Springer Nature Switzerland AG 2026
E. Cuevas et al., *Image Processing with Python*, Signals and Communication Technology, https://doi.org/10.1007/978-3-032-13285-7_2

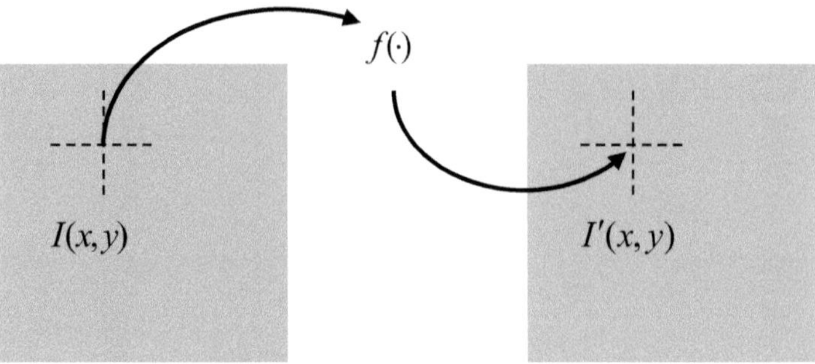

Fig. 2.1 Point operations where each output pixel derives exclusively from a function applied to its corresponding input pixel

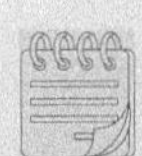

Pixel operations, also known as point operations, are fundamental in image processing and are widely applied in various domains due to their simplicity and efficiency. These operations involve modifying the intensity value of each pixel independently, without considering its neighbors. Common applications include brightness and contrast adjustment, where pixel values are scaled or shifted to enhance visibility, and image inversion, often used in medical imaging to highlight anatomical structures. Thresholding is another key application, converting grayscale images into binary images to separate objects from the background, which is essential in segmentation tasks such as object detection or character recognition. Pixel operations are also used for image normalization in machine learning preprocessing, masking specific regions of interest, and even in digital watermarking, where information is embedded directly into pixel values. Their versatility makes pixel operations a foundational tool in fields like medical diagnostics, remote sensing, industrial inspection, and computer vision.

2.1 Pixel-Wise Intensity Adjustment

2.1.1 Brightness and Contrast Variations

Image contrast refers to the relationship between varying intensity levels within the image. Brightness, or illumination [2], is linked to the overall distribution of these intensity values. When values cluster toward the lower end, the image appears darker; when they lean toward the higher end, the image seems brighter. As a practical

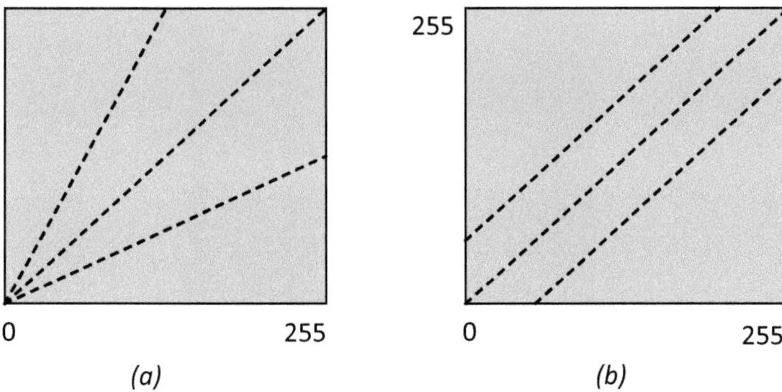

Fig. 2.2 Transformation mapping applied to output pixels $I'(x, y)$ when adjusting parameters: **a** c and **b** b in Eq. 2.4

illustration, consider enhancing an image's contrast by 50%. The method applies a consistent multiplier of 1.5 to every pixel in the image. Similarly, increasing brightness by 10 units can be achieved by adding 10 to each pixel. These homogeneous functions can, therefore, be expressed as follows:

$$f_c(I(x, y)) = I(x, y) \cdot 1.5 \quad \text{and} \quad f_h(I(x, y)) = I(x, y) + 10. \tag{2.3}$$

The transformation operator $f(\cdot)$, designed for contrast and luminance adjustment in digital images, is formally expressed as:

$$I(x, y) = f(x, y) = c \cdot I(x, y) + b. \tag{2.4}$$

The parameter c adjusts image contrast, while b controls brightness (or illumination). Figure 2.2 visually demonstrates the distinct effects produced by varying these two parameters.

Figure 2.3 visual outcomes of implementing the previously described uniform pixel operations on a sample image

2.1.2 Boundary Definition Through Pixel-Level Processing

When applying homogeneous operations [3], the resulting pixel value might go beyond the allowable range for 8-bit grayscale images, causing it to fall outside the interval between 0 and 255. As a result, any excess must be managed appropriately; it becomes essential to safeguard the program by incorporating the following conditional steps:

```
If (Ixy > 255)
Ixy = 255;
```

(a)

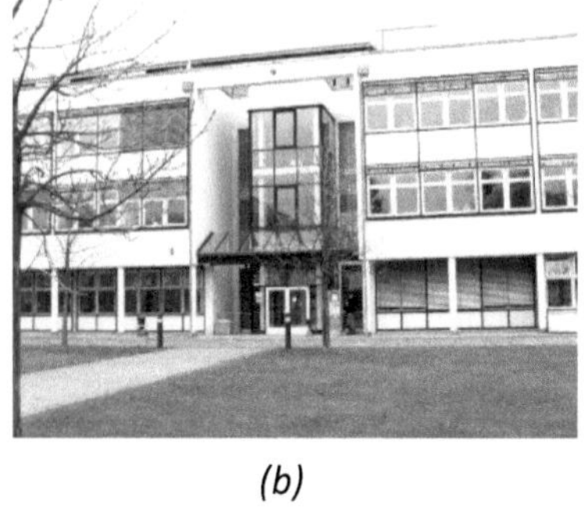

(b)

(c)

Fig. 2.3 Demonstration of uniform pixel transformations: **a** source image, **b** 50% contrast-adjusted version, and **c** brightness-enhanced image (+ 10 levels)

This operation will help remove any overflow caused by applying a homogeneous operation to the image. This adjustment is commonly known in the literature as "clamping." A similar issue arises when the resulting pixel value falls below the minimum threshold allowed in an 8-bit grayscale image, often due to reduced brightness levels, which may lead to negative values. Like the previous case, this issue can be prevented by securing the program with the following conditions:

```
If (Ixy < 0)
Ixy = 0;
```

2.1.3 *Pixel Inversion*

Image inversion is classified as a pixel-level operation [4], where the pixel's intensity is modified in the reverse direction (by applying a multiplication with -1). Additionally, a fixed intensity value is added to ensure the final result remains within the valid limits of the image format. For a pixel $p = I(x, y)$ within a defined intensity range, the inversion or complement process is expressed as follows:

Fig. 2.4 Outcome of the pixel inversion operation applied to an image. **a** Source image and **b** inverted result

$$f_{\text{inv}}(p) = p_{\text{max}} - p. \tag{2.5}$$

Figure 2.4 demonstrates the visual impact of applying a pixel complement operation to an image.

2.2 Pixel Operations in Python

Building upon the theoretical framework established in preceding sections, we now demonstrate practical Python implementations for pixel-level image processing.

2.2.1 Contrast and Illumination Change in Python

In Python, enhancing an image's contrast involves multiplying the image by a positive factor, which stretches the histogram and increases visual distinction. The following example assumes that A represents the original image, and applying a 50% contrast increase corresponds to multiplying A by 1.5 to produce the resulting image B.

```
>>B = A * 1.5;
```

To increase an image's brightness in Python, simply add a constant intensity value to each element in the pixel matrix. This shifts the histogram toward higher values, increasing illumination accordingly. The following line assumes we aim to brighten image A by 10 levels, so 10 is added to the matrix, and the result is stored in image B.

```
>>B = A + 10;
```

In Python, it is necessary to consider that if the multiplication exceeds the allowed data type for the image, values exceeding 255 will be truncated to the maximum value of 255.

CONTRAST CHANGE IN PYTHON

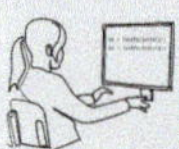

#CODE 2.1

```python
#The libraries necessary for image processing are loaded.

import cv2 as cv
import numpy as np
import matplotlib.pyplot as plt
from google.colab.patches import cv2_imshow

#The image is loaded for processing
img = cv.imread('2.jpg')
img = cv.resize(img,dsize=None,fx=0.15,fy=0.15)
cv2_imshow(img)
```

```python
#The image is transformed from color to grayscale
img = cv.cvtColor(img,cv.COLOR_BGR2GRAY)
cv2_imshow(img)
```

```
#Increased contrast
imgHC = img*1.5
cv2_imshow(imgHC)
```

```
#Increase contrast more
imgHC = img.copy()
imgHC = np.double(imgHC * 1.5)
np.putmask(imgHC, imgHC > 255, 255)
imgHC = np.uint8(imgHC)
cv2_imshow(imgHC)
```

```
#Another alternative to increase the contrast is to use for
cycles in python
rows,cols = np.shape(img)
imgHC = np.zeros([rows,cols])
for i in range(rows):
    for j in range(cols):
        pixel = np.double(img[i,j] * 1.5)
        pixel = img[i,j] * 1.5
        if pixel > 255:
            pixel = 255
        imgHC[i,j] = pixel
imgHC = np.uint8(imgHC)
cv2_imshow(imgHC)
```

```
#Code to apply less contrast
imgLC = img.copy()
imgLC = np.double(imgLC) * 0.5
np.putmask(imgLC, imgLC > 255, 255)
imgLC = np.uint8(imgLC)
cv2_imshow(imgLC)
```

```
#Applying less contrast with cyclo-for code
imgLC = np.zeros([rows,cols])
for i in range(rows):
    for j in range(cols):
        pixel = np.double(img[i,j]) * 0.5
        if pixel > 255:
          pixel = 255
        imgLC[i,j] = pixel
imgLC = np.uint8(imgLC)
cv2_imshow(imgLC)
```

```
#Code to apply brightness to the image
imgHB = img.copy()
imgHB = np.double(imgHB) + 100
np.putmask(imgHB, imgHB > 255, 255)
imgHB = np.uint8(imgHB)
cv2_imshow(imgHB)
```

```
#Applying brightness with cyclo-for code
imgHB = np.zeros([rows,cols])
for i in range(rows):
    for j in range(cols):
        pixel = np.double(img[i,j] + 100)
        if pixel > 255:
          pixel = 255
        imgHB[i,j] = pixel
imgHB = np.uint8(imgHB)
cv2_imshow(imgHB)
```

```
#Code to remove brightness in an image
imgLB = img.copy()
imgLB = np.double(imgLB) - 100
np.putmask(imgLB, imgLB < 0, 0)
imgLB = np.uint8(imgLB)
cv2_imshow(imgLB)

#Code to remove brightness with cyclo-for
imgLB = np.zeros([rows,cols])
for i in range(rows):
    for j in range(cols):
        pixel = np.double(img[i,j] - 100)
        if pixel < 0:
          pixel = 0
        imgLB[i,j] = pixel
imgLB = np.uint8(imgLB)
cv2_imshow(imgLB)
```

```
#Brightness and contrast control with convertScaleAbs function of
OpenCV library
#This function performs the following operation:

#img2 = abs(img*alpha + beta)
```

```python
#Where alpha is the contrast factor, and beta is the brightness

#After the operation, it takes the absolute value

#Contrast control
contrast = 1
# Brightness control
brightness = -100

img_openCV = img.copy()
img_openCV =
cv.convertScaleAbs(img_openCV,alpha=contrast,beta=brightness)
cv2_imshow(img_openCV)
```

```python
#The addWeighted function can also be used to control brightness
and contrast.
#This function does not have the same disadvantages as the
convertScaleAbs function since it does not calculate the absolute.
#The addWeighted function performs the following operation:

#dst = src1*alpha + src2*beta + gamma;
#addWeighted(src1, alpha, src2, beta, gamma[, dst[, dtype]]) ->
dst

#Calculates the weighted sum of two arrays
#Alpha control (0 to 127)
alpha = 1

#We set beta to zero so that it only considers one image.
beta = 0

# Brightness control (0-100)
gamma = -100
img_openCV2 = cv.addWeighted(img,alpha,img,beta, gamma)
cv2_imshow(img_openCV2)
```

2.2.2 *Complement of an Image with the Python*

COMPLEMENT OF AN IMAGE IN PYTHON

#CODE 2.2

```python
#The libraries necessary for image processing are loaded.
import cv2 as cv
import numpy as np
import matplotlib.pyplot as plt
from google.colab.patches import cv2_imshow

#The image is loaded for processing
img = cv.imread('2.jpg')
img = cv.resize(img,dsize=None,fx=0.15,fy=0.15)
cv2_imshow(img)
```

```
#The image is transformed from color to grayscale
img = cv.cvtColor(img,cv.COLOR_BGR2GRAY)
cv2_imshow(img)
```

```
#Complement operation
imgComp = 255 - img
cv2_imshow(imgComp)
```

2.2.3 Composite Pixel Transformations

Sometimes, pixel operations must be carried out where the resulting intensity value is influenced not only by the current pixel in one image but also by corresponding pixels in additional images. Figure 2.5 presents a visual example that illustrates this type of operation [5].

As illustrated in Fig. 2.5, the resulting pixel is produced by applying a function that combines two pixels from two separate images. The key aspect is that the pixels involved in the operation are aligned, meaning they occupy the same spatial location within their respective images.

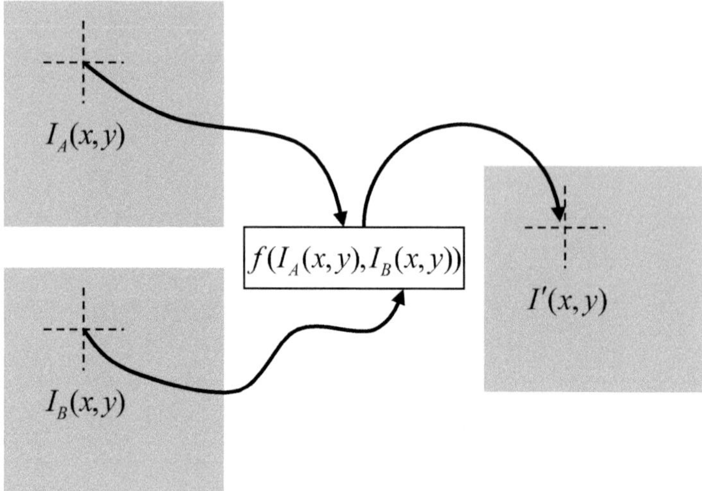

Fig. 2.5 Visualization of cross-image pixel processing, where output values derive from corresponding pixels in multiple source images while maintaining positional alignment

2.2.4 Boolean and Mathematical Computations

Element-wise image processing operations, including arithmetic (addition/subtraction) and logical (AND/OR) manipulations, are performed at the pixel level. These transformations can generate values exceeding the valid intensity range, potentially requiring subsequent normalization or rescaling procedures.

Operation addition

The resultant image from summing I_A and I_B is given by:

$$I'(x, y) = I_A(x, y) + I_B(x, y) \qquad (2.6)$$

A common application of this technique involves superimposing images to produce blended compositions, as illustrated in Fig. 2.6. Additionally, it can incorporate images containing scattered irregularities or dot patterns to either replicate natural noise or create artificial noise textures.

Subtraction operation

The dissimilarity between images I_A and I_B is mathematically expressed as follows:

$$I'(x, y) = I_A(x, y) - I_B(x, y). \qquad (2.7)$$

This technique is frequently employed in image segmentation and enhancement. Another key application of image subtraction is in detecting changes in position. By comparing two images captured at different times, T_1 and T_2, and calculating their

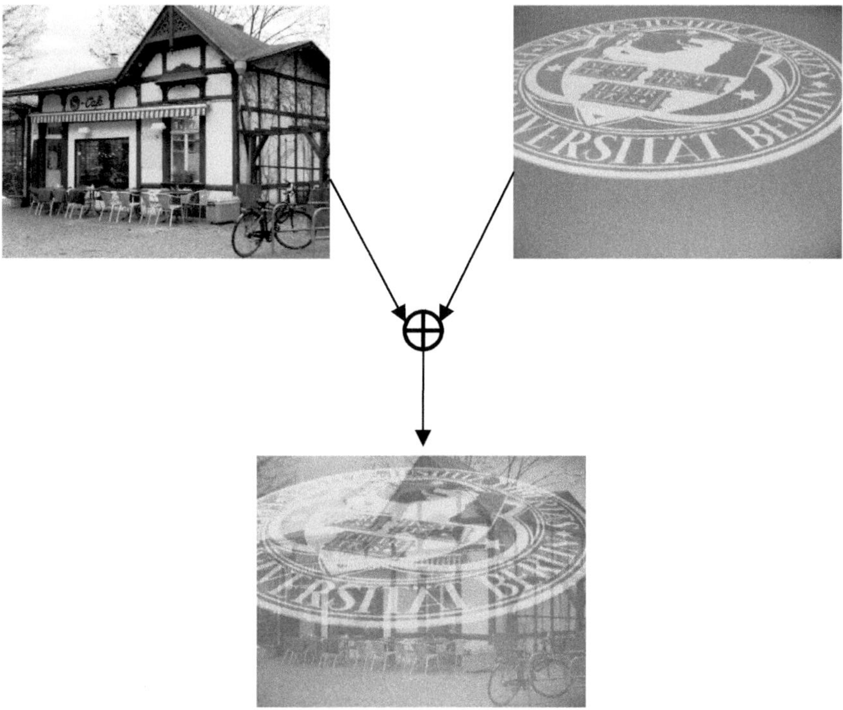

Fig. 2.6 Image composition through additive blending of two source images

difference, we can identify shifts in the pixels that belonged to an object that moved or altered its position. Figure 2.7 illustrates motion detection through pixel-wise subtraction of two images captured at distinct time intervals.

AND y OR

These procedures are executed on binary images following standard logic gate truth tables. The AND operation requires both input pixels to be 1 to yield a 1 in the output, while the OR operation produces a 1 if either pixel is 1. All other input combinations result in a 0 output value.

These operations employ localized processing, exclusively modifying pixels within defined block boundaries. Areas outside the target region remain unprocessed or are zeroed out.

2.2.5 *Alpha Mixing Operations*

The alpha blending or compositing technique merges two images, where each pixel in the resultant image is a weighted combination of the corresponding pixels from

Fig. 2.7 Motion detection through temporal image differencing, highlighting positional changes between frames captured at distinct times

both source images. Specifically, given two images, I_A and I_B, applying the alpha blending operation creates the output image I_R, where each pixel in I_R is a blend of the pixels from both I_A and I_B, as described by the following equation:

$$I_R(x, y) = (1 - \alpha) \cdot I_A(x, y) + \alpha \cdot I_B(x, y). \tag{2.8}$$

Through this process, two images can be blended, placing image I_A over the background image I_B. The degree to which I_A overlays I_B is adjusted using a transparency coefficient called α, which ranges from 0 to 1. Figure 2.8 illustrates how varying the α value affects the final composite result when the operation is applied to both images.

Fig. 2.8 Visualization of the alpha blending process. **a** Background image I_B, **b** foreground image I_A, **c** blended output I_R with $\alpha = 0.3$, and **d** blended output I_R with $\alpha = 0.7$

#CODE 2.3

```python
#The libraries necessary for image processing are loaded.
import cv2 as cv
import numpy as np
import matplotlib.pyplot as plt
from google.colab.patches import cv2_imshow

#Load image 1 for processing
img = cv.imread('2.jpg')

#Adjust image size
img = cv.resize(img,dsize=None,fx=0.15,fy=0.15)
cv2_imshow(img)
```

```python
#The image is transformed from color to grayscale
img = cv.cvtColor(img,cv.COLOR_BGR2GRAY)
cv2_imshow(img)
```

```python
#Load image 2 for processing
Img2 = cv.imread('7.jpg')

#Adjust image size, same as image 1
Img2 = cv.resize(img2,dsize=None,fx=0.15,fy=0.15)
cv2_imshow(img2)
```

```python
#The image is transformed from color to grayscale
Img2 = cv.cvtColor(img2,cv.COLOR_BGR2GRAY)
cv2_imshow(img2)
```

```python
#Combining the images with Alpha 0.3 for image 1 and Alpha 0.7
for the second image.

img3 = 0.3*img + 0.7*img2
cv2_imshow(img3)
```

```
#Now, the images are combined with Alpha 0.7 for image 1 and
Alpha 0.3 for the second image.
img3 = 0.7*img + 0.3*img2
cv2_imshow(img3)
```

```
# The addWeighted function can also be used to combine images.

#This function performs the following operation
#dst = src1*alpha + src2*beta + gamma;
#addWeighted(src1, alpha, src2, beta, gamma[, dst[, dtype]]) ->
dst
```

```
#Calculates the weighted sum of two arrays

#Alpha control (0 to 127)
alpha = 0.3
#Beta control (0 to 127)
beta = 0.7
#Brightness control (0-100)
gamma = 0

img_openCV2 = cv.addWeighted(img,alpha,img2,beta, gamma)
cv2_imshow(img_openCV2)
```

References

1. Liu, Z., Li, G., Mercier, G., He, Y., & Pan, Q. (2017). Change detection in heterogenous remote sensing images via homogeneous pixel transformation. *IEEE Transactions on Image Processing, 27*(4), 1822–1834.
2. Lin, H., Gao, J., Mei, Q., Zhang, G., He, Y., & Chen, X. (2017). Three-dimensional shape measurement technique for shiny surfaces by adaptive pixel-wise projection intensity adjustment. *Optics and Lasers in Engineering, 91*, 206–215.
3. Stockman, G., & Shapiro, L. G. (2001). Computer vision. Prentice Hall PTR.
4. Fairhurst, M. C. (1988). *Computer vision for robotic systems: An introduction.* Prentice Hall International (UK) Ltd.
5. Laganière, R. (2017). *OpenCV 3 computer vision application programming cookbook.* Packt Publishing Ltd.

Chapter 3
Geometric Operations in Images

There are operations on images that work on a pixel level or in the form of a filter. However, they share the commonality of altering the pixel's intensity value while maintaining the image geometry unchanged [1]. Images can be distorted through geometric operations that alter the positions of pixels. Such operations include rotation, translation, scaling, or inclination, as demonstrated in Fig. 3.1. Geometric operations in images are crucial for practical applications, especially in video games, graphical user interfaces, and mobile phone applications. Today, virtually all graphics applications feature zoom capabilities to highlight specific details within images. In the field of computer graphics, these operations are essential for representing textures, creating 3D environments, and facilitating real-time virtual representations of surroundings. However, despite their apparent simplicity, achieving good results with geometric operations can require a considerable amount of processing time, even on modern computers.

Formally, a spatial transformation can be described as a function that maps each point in the original image domain to a new position, such that:

$$I(x, y) \rightarrow I'(x', y'). \tag{3.1}$$

It implies that not only the pixel intensity, but also its spatial relation within the image matrix, is altered. The transformation T can be defined as:

$$T : \boldsymbol{R}^2 \rightarrow \boldsymbol{R}^2. \tag{3.2}$$

Every coordinate $x = (x, y)$ in the original image $I(x, y)$ defines its corresponding location $x' = (x', y')$ in the transformed image $I'(x', y')$. Such that:

$$x \rightarrow x' = T(x). \tag{3.3}$$

© The Author(s), under exclusive license to Springer Nature Switzerland AG 2026
E. Cuevas et al., *Image Processing with Python*, Signals and Communication
Technology, https://doi.org/10.1007/978-3-032-13285-7_3

Fig. 3.1 Various examples
of geometric transformations
to be explored throughout
this chapter

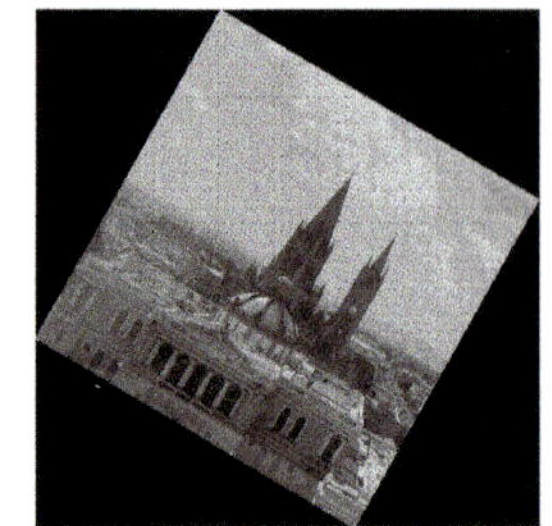

As indicated by the preceding equations, both original and transformed coordinates are treated as continuous points within the real number plane $R \times R$. This mathematical abstraction simplifies the definition of geometric transformations. However, in practical applications, images are composed of a discrete grid of pixels, meaning that actual coordinates fall within integer-valued positions on this grid. Consequently, when a transformation is applied, the resulting coordinates rarely align perfectly with existing pixel locations.

This discrepancy introduces uncertainty, as the transformed point will often lie between two or more-pixel positions. Since digital images cannot represent values at fractional coordinates directly, an additional step is necessary to resolve this issue. The standard approach involves interpolation, a method for estimating pixel values at non-integer positions by considering surrounding known values. Interpolation plays a crucial role in implementing geometric transformations, ensuring visual continuity and minimizing artifacts when mapping transformed images back onto a discrete pixel grid.

3.1 Coordinate Transformation

A geometric transformation [2], as expressed in Eq. 3.3, represents a function that maps an original coordinate to a new location in space. This transformation can be broken down into two separate functions, one for each axis:

$$x' = T_x(x, y) \text{ and } y' = T_y(x, y). \tag{3.4}$$

3.1.1 Basic Transformations

The most common elementary transformations include translation, scaling, inclination, and rotation. Each of these operations alters the spatial arrangement of pixels in a distinct manner.

Translation

Translation shifts the entire image in space by applying a displacement vector (d_x, d_y) to each pixel:

$$T_x : x' = x + d_x \text{ and } T_y : y' = y + d_y. \tag{3.5}$$

Expressed in matrix form:

$$\begin{bmatrix} x' \\ y' \end{bmatrix} = \begin{bmatrix} x \\ y \end{bmatrix} + \begin{bmatrix} d_x \\ d_y \end{bmatrix}.$$

Scaling

Scaling adjusts the image size along the horizontal $x(s_x)$ or vertical $y(s_y)$ axis. The transformation enlarges or compresses the image based on the scaling factors:

$$\begin{aligned} T_x : x' = x \cdot s_x \\ T_y : y' = y \cdot s_y \end{aligned} \tag{3.6}$$

In matrix form:

$$\begin{bmatrix} x' \\ y' \end{bmatrix} = \begin{bmatrix} s_x & 0 \\ 0 & s_y \end{bmatrix} \begin{bmatrix} x \\ y \end{bmatrix}.$$

Inclination

Skew the image by shifting one axis in proportion to the other. Horizontal shearing uses coefficient $x(b_x)$ while vertical shearing uses $y(b_y)$:

$$T_x : x' = x + b_x \cdot y$$
$$T_y : y' = y + b_y \cdot x. \qquad (3.7)$$

The expression is presented in matrix form:

$$\begin{bmatrix} x' \\ y' \end{bmatrix} = \begin{bmatrix} 1 & b_x \\ b_y & 1 \end{bmatrix}\begin{bmatrix} x \\ y \end{bmatrix}.$$

Rotation

Rotation repositions each pixel based on a given angle α, typically using the image's center as the pivot point. The formulas are:

$$T_x : x' = x \cdot \cos(\alpha) + y \cdot \sin(\alpha)$$
$$T_y : y' = -x \cdot \sin(\alpha) + y \cdot \cos(\alpha). \qquad (3.8)$$

Matrix representation:

$$\begin{bmatrix} x' \\ y' \end{bmatrix} = \begin{bmatrix} \cos(\alpha) & \sin(\alpha) \\ -\sin(\alpha) & \cos(\alpha) \end{bmatrix}\begin{bmatrix} x \\ y \end{bmatrix}.$$

These simple geometric transformations form the foundation of many image processing and computer graphics operations. Their visual effects are typically illustrated in diagrams, as shown in Fig. 3.2.

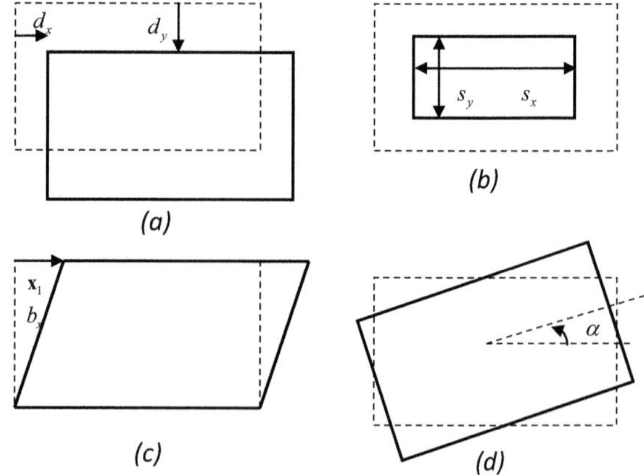

Fig. 3.2 Basic types of geometric transformations: These exhibit the following operations: **a** Translation: shifts the entire image in the x and y axes; **b** rescaling: modifies size of the image by expanding or contracting its dimensions; **c** inclination: tilts the image shape by slanting it horizontally or vertically; **d** rotation: turns the image around a fixed center point to a specified angle

3.1.2 *Homogeneous Coordinates*

The transformations described in basic operations, such as translation, scaling, inclination, and rotation, belong to a broader group known as affine transformations. When
multiple transformations need to be combined [2], it becomes helpful to represent
them using matrix operations. A streamlined method to handle such combinations is
by employing homogeneous coordinates.

In this system, a 2D coordinate vector is extended by adding a third component,
typically denoted as h. This transforms the point from a two-dimensional vector
(x, y) into a three-dimensional vector:

$$x = \begin{bmatrix} x \\ y \end{bmatrix} \rightarrow \hat{x} = \begin{bmatrix} \hat{x} \\ \hat{y} \\ h \end{bmatrix} = \begin{bmatrix} hx \\ hy \\ h \end{bmatrix}. \tag{3.9}$$

With this representation, any point in the 2D Cartesian plane can be expressed
in homogeneous form. When the h component is nonzero, the Cartesian coordinates
can be recovered using:

$$x = \frac{\hat{x}}{h} \text{ and } y = \frac{\hat{y}}{h}. \tag{3.10}$$

One important aspect of homogeneous coordinates is that a single 2D point can
have an infinite number of equivalent homogeneous representations. For instance,
the vectors $\hat{x}_1 = (2, 1, 1)$, $\hat{x}_1 = (4, 2, 2)$, and $\hat{x}_1 = (20, 10, 10)$ all correspond to the
same Cartesian point $(2, 1)$, since each one reduces to the exact coordinates when
divided by their respective h values.

This technique not only simplifies the mathematical treatment of combined transformations but also enables the use of unified matrix operations for both affine and
projective mappings.

3.1.3 *Affine Transformation (Triangle Transformation)*

Using homogeneous coordinates provides a robust framework for combining
multiple geometric operations, such as translation, scaling, rotation, and inclination,
into a single matrix expression. With this approach, a 2D point can be transformed
as follows:

$$\begin{bmatrix} \hat{x}' \\ \hat{y}' \\ \hat{h}' \end{bmatrix} = \begin{bmatrix} x' \\ y' \\ 1 \end{bmatrix} = \begin{bmatrix} a_{11} & a_{12} & a_{13} \\ a_{21} & a_{22} & a_{23} \\ 0 & 0 & 1 \end{bmatrix} \begin{bmatrix} x \\ y \\ 1 \end{bmatrix}. \tag{3.11}$$

This formulation is known as an affine transformation [3], characterized by six parameters $(a_{11}, a_{12}, a_{13}, a_{21}, a_{22}, a_{23})$. This values a_{13} and a_{23} define translation, while the other four parameters are responsible for scaling, inclination, and rotation. Affine transformations are linear mappings that preserve parallelism, meaning they convert straight lines into other straight lines.

A key feature of affine transformations is their ability to map basic shapes predictably: triangles remain triangles, and rectangles become parallelograms. While transformations may alter angles and lengths, the proportional spacing of points along a line remains unchanged, preserving the overall geometric relationships in the modified image. This concept is demonstrated in Fig. 3.3.

Computing the Transformation Parameters

To solve for the six parameters of the affine transformation matrix, three pairs of matching points are sufficient: (x_1, x_1'), (x_2, x_2'), and (x_3, x_3'), where each $x_i = (x_i, y_i)$ denotes a point in the original image, and $x_i' = (x_i', y_i')$ denotes its counterpart in the transformed image [4].

Solving for the affine parameters involves setting up and solving a linear system based on these point correspondences, which ensures that the transformation matrix correctly maps each original point to its new location [5].

$$x_1' = a_{11} \cdot x_1 + a_{12} \cdot y_1 + a_{13} \quad y_1' = a_{21} \cdot x_1 + a_{22} \cdot y_1 + a_{23}$$
$$x_2' = a_{11} \cdot x_2 + a_{12} \cdot y_2 + a_{13} \quad y_2' = a_{21} \cdot x_2 + a_{22} \cdot y_2 + a_{23}$$
$$x_3' = a_{11} \cdot x_3 + a_{12} \cdot y_3 + a_{13} \quad y_3' = a_{21} \cdot x_3 + a_{22} \cdot y_3 + a_{23}. \tag{3.12}$$

For the system of equations to yield a valid solution, the three point pairs—(x_1, x_1'), (x_2, x_2'), and (x_3, x_3')—must be linearly independent, meaning the original points must not all lie on the same straight line. If this condition is met, the affine transformation matrix can be uniquely determined. Solving this system provides the transformation parameters as expressed in Eq. 3.13 [6].

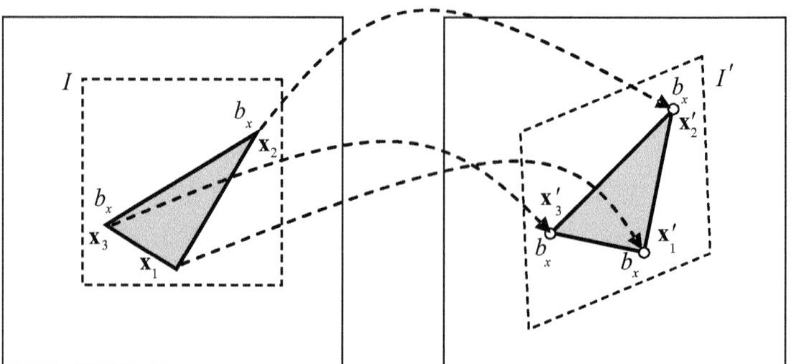

Fig. 3.3 Affine transformation, showing how defining the transformation of just three points is sufficient to determine the full behavior of the mapping. It is why affine transformations are sometimes referred to as "triangle-preserving" transformations

$$a_{11} = \frac{1}{F} \cdot \left[y_1\left(x_2' - x_3'\right) + y_2\left(x_3' - x_1'\right) + y_3\left(x_1' - x_2'\right) \right].$$

$$a_{12} = \frac{1}{F} \cdot \left[x_1\left(x_3' - x_2'\right) + x_2\left(x_1' - x_3'\right) + x_3\left(x_2' - x_1'\right) \right].$$

$$a_{21} = \frac{1}{F} \cdot \left[y_1\left(y_2' - y_3'\right) + y_2\left(y_3' - y_1'\right) + y_3\left(y_1' - y_2'\right) \right].$$

$$a_{22} = \frac{1}{F} \cdot \left[x_1\left(y_3' - y_2'\right) + x_2\left(y_1' - y_3'\right) + x_3\left(y_2' - y_1'\right) \right].$$

$$a_{13} = \frac{1}{F} \cdot \left[x_1\left(y_3 x_2' - y_2 x_3'\right) + x_2\left(y_1 x_3' - y_3 x_1'\right) + x_3\left(y_2 x_1' - y_1 x_2'\right) \right].$$

$$a_{23} = \frac{1}{F} \cdot \left[x_1\left(y_3 y_2' - y_2 y_3'\right) + x_2\left(y_1 y_3' - y_3 y_1'\right) + x_3\left(y_2 y_1' - y_1 y_2'\right) \right].$$

$$F = x_1\left(y_3 - y_2\right) + x_2\left(y_1 - y_3\right) + x_3\left(y_2 - y_1\right). \tag{3.13}$$

The Inverse of the Affine Transformation

It is essential when reversing geometric alterations applied to an image, such as mapping transformed coordinates back to their original positions. This inverse is computed by inverting the matrix used in the affine transformation (Eq. 3.11), it would be:

$$\begin{bmatrix} x \\ y \\ 1 \end{bmatrix} = \begin{bmatrix} a_{11} & a_{12} & a_{13} \\ a_{21} & a_{22} & a_{23} \\ 0 & 0 & 1 \end{bmatrix}^{-1} \begin{bmatrix} x' \\ y' \\ 1 \end{bmatrix}$$

$$= \frac{1}{a_{11}a_{22} - a_{12}a_{21}} \begin{bmatrix} a_{22} & -a_{12} & a_{12}a_{23} - a_{13}a_{22} \\ -a_{21} & a_{11} & a_{12}a_{21} - a_{11}a_{23} \\ 0 & 0 & a_{11}a_{22} - a_{12}a_{21} \end{bmatrix} \cdot \begin{bmatrix} x' \\ y' \\ 1 \end{bmatrix}. \tag{3.14}$$

Like the forward transformation, the parameters of the inverse transformation can be derived by specifying three pairs of matching points between the transformed image and the original. These values are substituted into the previously derived Eq. (3.13) to determine the inverse matrix. Figure 3.4 demonstrates the result of an affine transformation, where the original image is altered based on specific point mappings: $x_1 = (400, 300)$, $x_1' = (200, 280)$, $x_2 = (250, 20)$, $x_2' = (255, 18)$, $x_3 = (100, 100)$, and $x_3' = (120, 112)$. The resulting image shows the geometric transformation effect on the spatial arrangement of the original content.

(a)   (b)

Fig. 3.4 Effect of an affine transformation applied to an image. **a** Original image, **b** image after applying the transformation. The mapping was defined using three pairs of corresponding points: $x_1 = (400, 300)$, $x_1' = (200, 280)$, $x_2 = (250, 20)$, $x_2' = (255, 18)$, $x_3 = (100, 100)$, and $x_3' = (120, 112)$

3.2   Implementing Geometric Transformations in Python

This section introduces a simple test application written in Python to illustrate the implementation of the geometric transformations discussed earlier. The program includes functions to perform basic operations such as translation, scaling, inclination, and rotation. These implementations provide a hands-on approach to understanding how image data can be manipulated through mathematical transformations, using Python image processing libraries [7].

3.2.1   Translation of an Image with Python

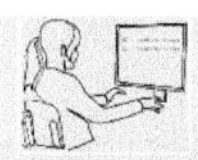

#Code 3.1 Translation

```python
#The libraries necessary for image processing are loaded.

import cv2 as cv
import numpy as np
import matplotlib.pyplot as plt
from google.colab.patches import cv2_imshow

#The image is loaded for processing
img = cv.imread('fig0301.png')

#The image is transformed from color to grayscale
img=cv.cvtColor(img,cv.COLOR_BGR2GRAY)
cv2_imshow(img)
```

```python
# Simple transformation, displacement
rows, cols = np.shape(img)
vecSize = rows*cols
tx = 150
ty = 120
T = [[1,0,0],[0,1,0],[tx,ty,1]]
v = np.zeros(vecSize)
w = np.zeros(vecSize)
k = 0
for x in range(rows):
  for y in range(cols):
    v[k] = round(x * T[0][0] + y * T[1][0] + T[2][0])
    w[k] = round(x * T[0][1] + y * T[1][1] + T[2][1])
    k = k+1
```

```
max_v = np.max(v)
max_w = np.max(w)
imgdisplac = np.zeros([int(max_v), int(max_w)])
k = 0
for x in range(rows):
  for y in range(cols):
    imgdisplac[int(v[k])-1,int(w[k])-1] = img[x,y]
    k = k+1
```

```
# Show the image after applying the translation transformation
cv2_imshow(imgdisplac)
```

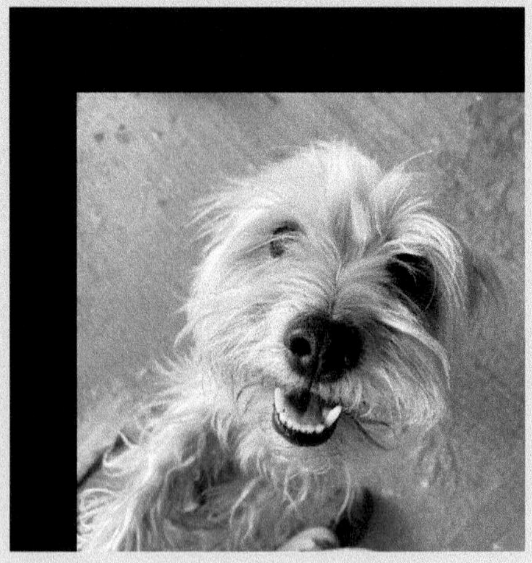

3.2.2 *Scaling of an Image with Python*

```
        %Code 3.2 Scaling

#The libraries necessary for image processing are loaded.

import cv2 as cv
import numpy as np
import matplotlib.pyplot as plt
from google.colab.patches import cv2_imshow

#The image is loaded for processing
img = cv.imread('fig0302.png')

#The image is transformed from color to grayscale
img=cv.cvtColor(img,cv.COLOR_BGR2GRAY)
cv2_imshow(img)
```

```
# simple transformation, scaling
fx = 0.5
fy = 0.5
k = 0
T = [[fx,0,0],[0,fy,0],[0,0,1]]
v = np.zeros(vecSize)
w = np.zeros(vecSize)
for x in range(rows):
  for y in range(cols):
    v[k] = x * T[0][0] + y * T[1][0] + T[2][0]
    w[k] = x * T[0][1] + y * T[1][1] + T[2][1]
    k = k+1
max_v = int(np.max(v))
```

```
max_w = int(np.max(w))
imgScale = np.zeros([max_v, max_w])
k = 0
for x in range(rows):
  for y in range(cols):
    imgScale[int(v[k]-1),int(w[k])-1] = img[x,y]
    k = k+1

# Show the image after applying the scaling transformation
cv2_imshow(imgScale)
```

```
fx = 1.2
fy = 1.2
k = 0
T= [[fx, 0, 0],[0, fy, 0], [0, 0, 1]]
v = np.zeros(vecSize)
w = np.zeros(vecSize)
for x in range(rows):
  for y in range(cols):
    v[k] = x * T[0][0] + y * T[1][0] + T[2][0]
    w[k] = x * T[0][1] + y * T[1][1] + T[2][1]
    k = k+1
max_v = int(np.max(v))
max_w = int(np.max(w))
imgScale = np.zeros([max_v, max_w])
k = 0
for x in range(rows):
  for y in range(cols):
    imgScale[int(v[k])-1,int(w[k])-1] = img[x,y]
    k = k+1

# Show the image after applying the translation transformation
cv2_imshow(imgScale)
```

3.2.3   Inclination of an Image with Python

```
%Code 3.3 inclination

#The libraries necessary for image processing are loaded.
import cv2 as cv
import numpy as np
import matplotlib.pyplot as plt
from google.colab.patches import cv2_imshow

# for image display
img = cv.imread('fig0303.png')
img = cv.cvtColor(img,cv.COLOR_BGR2GRAY)
cv2_imshow(img)

rows, cols = np.shape(img)
vecSize = rows*cols
```

```
#Simple transformation, inclination horizontal
Ih = 0.2

#Ranges from 0 to 1, Ih is the ratio in which the image is
stretched horizontally.
T= [[1, Ih, 0],[0, 1, 0], [0, 0, 1]]
v = np.zeros(vecSize)
w = np.zeros(vecSize)
k = 0
for x in range(rows):
  for y in range(cols):
    v[k] = round(x * T[0][0] + y * T[1][0] + T[2][0])
    w[k] = round(x * T[0][1] + y * T[1][1] + T[2][1])
```

```
      k = k+1
max_v = np.max(v)
max_w = np.max(w)
imgIncHor = np.zeros([int(max_v), int(max_w)])
k = 0
for x in range(rows):
  for y in range(cols):
    imgIncHor[int(v[k])-1,int(w[k])-1] = img[x,y]
    k = k+1
cv2_imshow(imgIncHor)
```

```python
#Simple transformation, inclination vertically
Iv = 0.2

# Ranges from 0 to 1, Iv is the proportion by which the image is
stretched vertically
T= [[1, 0, 0],[Iv, 1, 0], [0, 0, 1]]
v = np.zeros(vecSize)
w = np.zeros(vecSize)
k = 0
for x in range(rows):
  for y in range(cols):
    v[k] = round(x * T[0][0] + y * T[1][0] + T[2][0])
    w[k] = round(x * T[0][1] + y * T[1][1] + T[2][1])
    k = k+1
max_v = np.max(v)
max_w = np.max(w)
imgIncVer = np.zeros([int(max_v), int(max_w)])
k = 0
for x in range(rows):
  for y in range(cols):
    imgIncVer[int(v[k])-1,int(w[k])-1] = img[x,y]
```

```python
    k = k+1
cv2_imshow(imgIncVer)
```

3.2.4   Rotation of an Image with Python

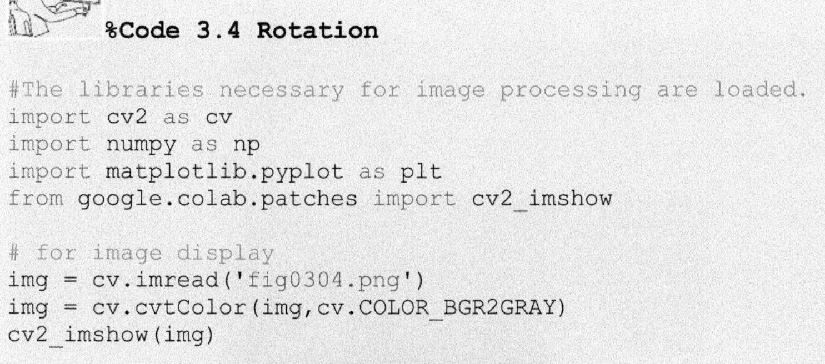

```
       %Code 3.4 Rotation

#The libraries necessary for image processing are loaded.
import cv2 as cv
import numpy as np
import matplotlib.pyplot as plt
from google.colab.patches import cv2_imshow

# for image display
img = cv.imread('fig0304.png')
img = cv.cvtColor(img,cv.COLOR_BGR2GRAY)
cv2_imshow(img)
```

```
# simple transformation, rotation
rows, cols = np.shape(img)
vecSize = rows*cols
k = 0
degreess = -30
radians = degreess * np.pi / 180
T= [[np.cos(radians), np.sin(radians), 0],[-np.sin(radians),
np.cos(radians), 0], [0, 0, 1]]
v = np.zeros(vecSize)
w = np.zeros(vecSize)
for x in range(rows):
  for y in range(cols):
    v[k] = round(x * T[0][0] + y * T[1][0] + T[2][0])
    w[k] = round(x * T[0][1] + y * T[1][1] + T[2][1])
    k = k+1
min_v = np.min(v)
```

```python
min_w = np.min(w)
v = v - min_v
w = w - min_w
max_v = np.max(v)
max_w = np.max(w)
imgRota = np.zeros([int(max_v), int(max_w)])
k = 0
for x in range(rows):
  for y in range(cols):
    imgRota[int(v[k])-1,int(w[k])-1] = img[x,y]
    k = k+1
cv2_imshow(imgRota)
```

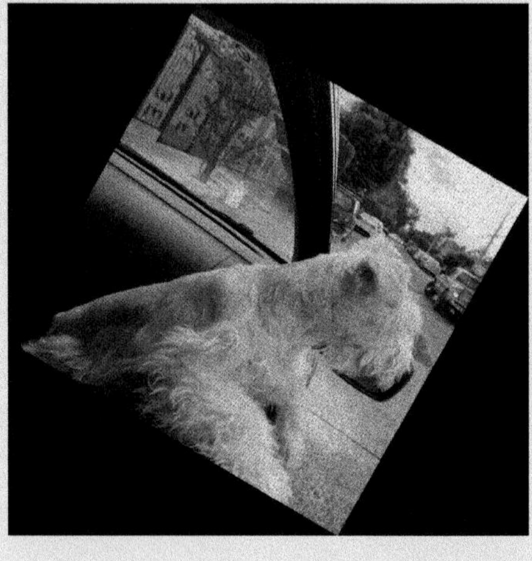

References

1. Gonzalez, R. C., & Woods, R. E. (2018). *Digital image processing* (4th ed.). Pearson.
2. Szeliski, R. (2022). *Computer vision: Algorithms and applications* (2nd ed.). Springer.
3. Sonka, M., Hlavac, V., & Boyle, R. (2014). *Image processing, analysis, and machine vision* (4th ed.). Cengage Learning.
4. Burger, W., & Burge, M. J. (2016). *Digital image processing: An algorithmic introduction using Java* (2nd ed.). Springer.
5. Pratt, W. K. (2007). Digital image processing: PIKS inside (4th ed.). Wiley-Interscience.
6. Wolberg, G. (1990). *Digital image warping.* IEEE Computer Society Press.
7. Bradski, G., & Kaehler, A. (2008). *Learning OpenCV: Computer vision with the OpenCV library.* O'Reilly Media.

Chapter 4
Histograms

In image processing, a histogram is a graphical representation that shows the distribution of pixel intensity values within an image. It plots the number of pixels for each possible intensity level, typically ranging from 0 (black) to 255 (white) for grayscale images. This visual summary is essential for understanding the contrast, brightness, and overall exposure of an image. Histograms are widely used in tasks such as image enhancement, where techniques like histogram equalization improve contrast; in segmentation, where thresholding is guided by intensity distributions; and in feature extraction for object recognition and classification. By analyzing the histogram, practitioners can make informed adjustments to improve image quality or prepare images for further processing.

4.1 Definition of a Histogram

Histograms [1] represent the distribution of intensity values in an image by showing how often each value occurs. In the most straightforward case, histograms are easiest to grasp with grayscale images. For example, take a grayscale image $I(u, v)$ where intensity values range from 0 to $K - 1$. The histogram H for this image will have exactly K values. In a standard 8-bit grayscale image, this results in $H = 2^8 = 256$ intensity levels. Each element in the histogram, indicated as $h(i) = a$, signifies the count of pixels in the image III with an intensity value of iii, for all $0 \leq i < K$. This formal definition provides a quantitative representation of how pixel intensities are distributed throughout the image. This condition can be defined as (Fig. 4.1):

$$h(i) = \mathrm{card}^2\{(u, v)|I(u, v) = i\}. \tag{4.1}$$

© The Author(s), under exclusive license to Springer Nature Switzerland AG 2026
E. Cuevas et al., *Image Processing with Python*, Signals and Communication
Technology, https://doi.org/10.1007/978-3-032-13285-7_4

 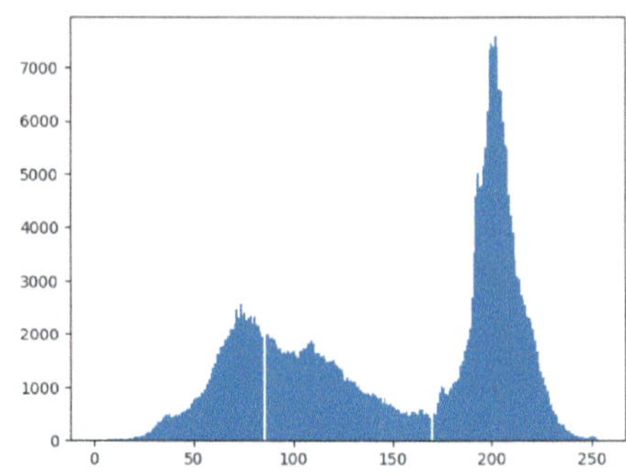

Fig. 4.1 Grayscale 8-bit image and corresponding histogram

$h(0)$ corresponds to the count of pixels with a value of 0. On the other hand, $h(1)$ indicates the count of pixels with a value of 1, and so on, until $h(255)$, which denotes the number of white pixels (those with the highest intensity value) in the image. This histogram calculation results in a one-dimensional vector h, which has a length of K, as illustrated in Fig. 4.2, where K equals 16.

While histograms are powerful tools for analyzing the intensity distribution of an image, they have a significant limitation: they do not retain any information about the spatial arrangement or origin of the pixels. This means that, although a histogram

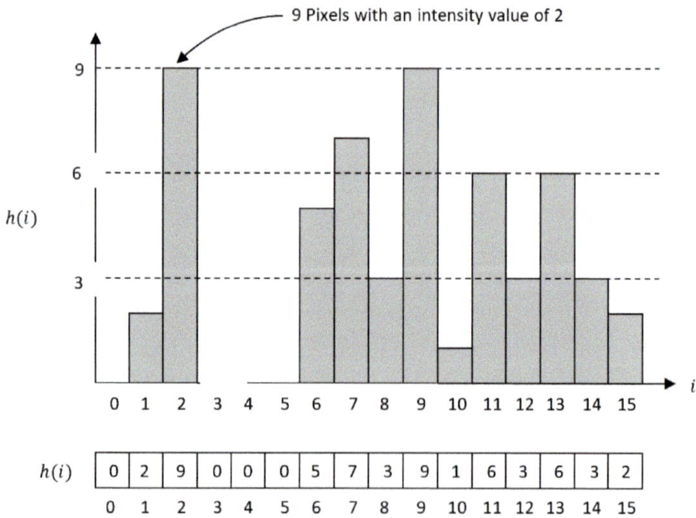

Fig. 4.2 Histogram is represented as a vector with 16 possible intensity values, where each $i = 1, \ldots, 15$ corresponds to a specific intensity level. The number stored at each index indicates how often that intensity occurs in the image. For instance, if the value at position 2 in the vector is 9, this means that the intensity level 2 appears 9 times in the image

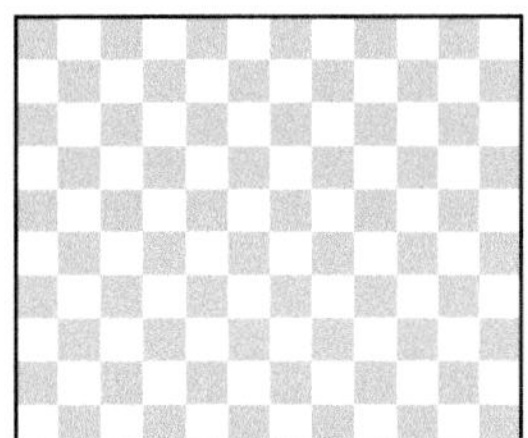

Fig. 4.3 Identical histograms from visually distinct images

tells us how many pixels have a certain intensity value, it does not indicate where those pixels are located within the image. As a result, it is impossible to reconstruct the original image solely from its histogram. This concept is illustrated in Fig. 4.3, where three visually distinct images produce identical histograms, demonstrating that different spatial configurations can yield the same intensity distribution. Despite this limitation, histograms still reveal critical characteristics of an image, such as contrast and dynamic range. These features are essential for understanding image quality and are often influenced by imperfections during the image capture process, such as poor lighting or sensor limitations. Such issues can obscure important visual information, making it necessary to apply post-processing techniques—like contrast enhancement or histogram equalization—to improve the clarity and usefulness of the image. Therefore, while the histogram cannot describe the spatial structure of an image, it remains a key diagnostic tool for identifying and correcting problems that affect the visual and analytical interpretation of image data.

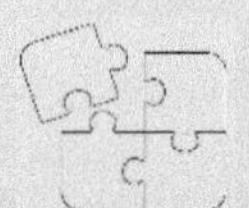

Histograms are versatile tools used across many fields beyond image processing, serving as fundamental instruments for analyzing data distributions, detecting patterns, and supporting decision-making. In statistics and data analysis, histograms provide an intuitive way to visualize the frequency distribution of numerical data, making them essential for understanding variability, detecting skewness, and identifying outliers. They are commonly used in quality control and manufacturing, where histograms help monitor production consistency and identify deviations from expected performance.

In machine learning, histograms play a critical role in feature extraction and data preprocessing. For instance, they are used in methods like Histogram of Oriented Gradients (HOG) for object detection, or in discretizing continuous features into bins before applying algorithms such as decision trees or Naive Bayes classifiers. In finance, histograms help assess the distribution of returns, detect volatility, and compare the behavior of different assets over time.

Histograms are also valuable in medical research and bioinformatics for visualizing distributions of biological measurements (e.g., blood pressure, gene expression levels) and in physics and astronomy for interpreting experimental results and observing distributions of particle counts, energy levels, or brightness.

Moreover, in environmental science, histograms are used to study weather patterns, pollution levels, or temperature distributions over time. Across all these disciplines, histograms are essential because they offer a simple yet powerful graphical representation of data, enabling users to explore and communicate insights clearly and effectively.

4.2 Fundamentals of Image Acquisition

Histograms reveal key characteristics of an image, such as contrast and dynamic range [2], by illustrating how pixel intensities are distributed across the image. A well-spread histogram typically indicates good contrast, while a narrow or skewed histogram may suggest low contrast or over-/underexposure. These features are crucial for assessing the overall visual quality of an image. Additionally, histograms can expose problems introduced during image capture, such as poor lighting, sensor limitations, or noise, which may result in loss of detail or distortion in specific intensity ranges. These capture-related issues directly impact subsequent image processing steps—such as enhancement, segmentation, or recognition—since they rely on accurate and well-distributed intensity information. Therefore, understanding and interpreting histograms is essential not only for image evaluation but also for preparing images for further analysis or correction.

4.2.1 Illumination Issues

Illumination errors in an image can be detected by analyzing its histogram, as these issues often cause certain regions of the intensity scale to remain unused. Specifically, when either the darkest or brightest parts of the intensity range are absent, it indicates that the image lacks proper exposure—either being too dark (underexposed) or too bright (overexposed). This results in a histogram where the pixel intensity values are concentrated in only a portion of the scale, leaving the initial or final segments empty. Such uneven distributions reflect a limited dynamic range and poor contrast, which can obscure image details and reduce visual quality. Figure 4.4 illustrates this concept by presenting images affected by various types of illumination, showing how their corresponding histograms reveal the underlying exposure problems. Identifying these

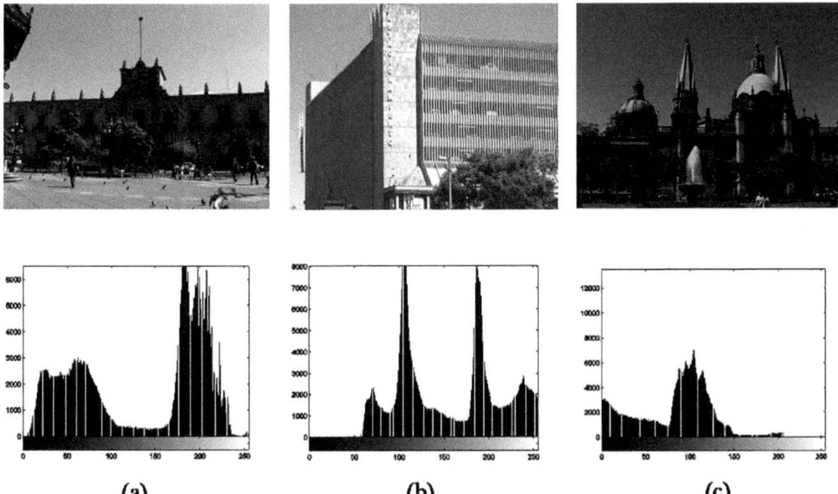

(a)   (b)   (c)

Fig. 4.4 Figures illustrate how histograms can reveal illumination issues in images: **a** an image with proper lighting, **b** an image that is overexposed, and **c** an image that is underexposed

patterns is essential for correcting illumination issues and improving the effectiveness of subsequent image processing tasks.

4.2.2 Contrast Properties

Contrast [3] refers to the range of intensity values utilized within an image and can be understood as the difference between the maximum and minimum pixel intensity values present. In essence, it measures how distinctly light and dark areas are represented. An image with full contrast spans the entire available intensity range, from the minimum $a_{\min} = 0$ (black) to the maximum $a_{\max} = K - 1$ (white), where K represents the number of possible intensity levels, such as 256 for an 8-bit image. When an image uses this full range, it generally appears well-defined and visually rich. In contrast, images with poor contrast have intensity values confined to a narrow segment of the available range, resulting in a flat or washed-out appearance. This difference is easily identifiable by inspecting the image's histogram: a wide, evenly distributed histogram indicates high contrast, while a narrow, clustered histogram signifies low contrast. Figure 4.5 demonstrates this by showing images with varying levels of contrast alongside their corresponding histograms, clearly illustrating the relationship between contrast and intensity distribution.

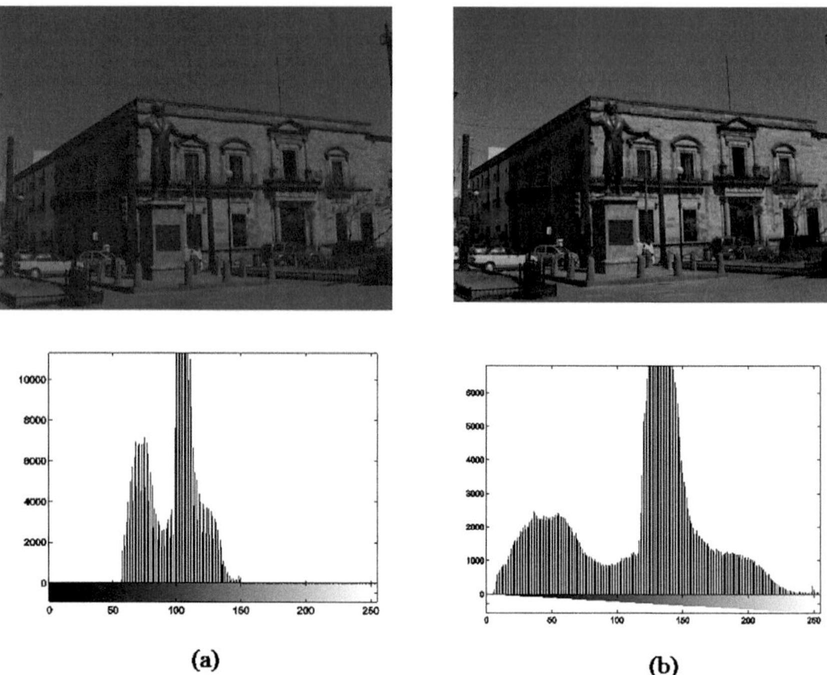

(a)   (b)

Fig. 4.5 Figures display images with varying contrast levels and their corresponding histograms: **a** an image with low contrast and **b** an image with standard or normal contrast

4.2.3   Concept of Dynamics

The term *dynamics* in image processing refers to the range and distribution of intensity values actually used by the pixels in an image. Ideally, a high-quality image should utilize the full available intensity range, from the minimum to the maximum possible values (e.g., 0–255 in an 8-bit image), ensuring a rich and detailed representation. This condition indicates that the image has reached its maximum dynamic range, meaning all intensity levels within that range are present. However, in many cases, images may only use a limited portion of the full intensity scale, where the minimum intensity $a_{\min} > 0$ and the maximum intensity $a_{\max} < 256$. Even in these cases, if all values between $a_{\min}$ and $a_{\max}$ are represented, the image is considered to have good dynamics within that narrower range. Figure 4.6 illustrates this concept by showing how different images may occupy varying portions of the intensity scale, and how the presence or absence of certain values impacts the image's dynamic range. Proper dynamics are essential for ensuring visual richness and maximizing the information content in an image.

While an image can exhibit high contrast as long as there is a significant difference between the darkest and brightest pixel values—even if these do not reach the absolute limits of the intensity scale—the same does not apply to image dynamics. High

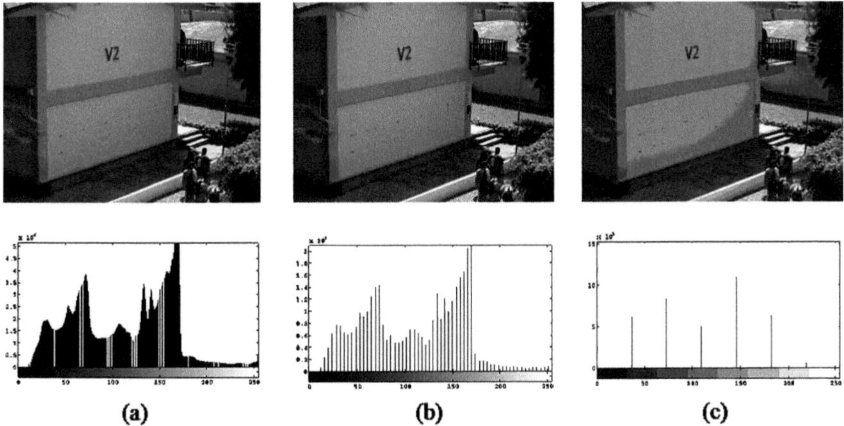

Fig. 4.6 Variations in image dynamics and their impact on the histogram are shown in the figures: **a** high dynamic range, **b** reduced dynamics using 64 intensity levels, and **c** very limited dynamics with just 8 intensity levels

dynamics require that a wide and continuous range of intensity values be used, and this cannot be artificially increased without interpolation, which does not add real information. Having high dynamics is advantageous because it preserves more detail and tonal variation, minimizing the risk of losing image quality during further processing steps such as enhancement, segmentation, or compression. To support this, digital cameras and professional scanners are typically designed to capture images with higher bit-depths—commonly 12–14 bits—far beyond the standard 8-bit (256-level) display resolution. This ensures that the captured image contains a richer set of intensity values, which can then be adjusted, compressed, or displayed as needed, without compromising the original quality or losing important visual information.

Lighting and contrast play a fundamental role in the quality and clarity of an image during the capture process. Inadequate or uneven lighting can lead to a range of problems, such as underexposure, where the image appears too dark, or overexposure, where details are lost in excessively bright regions. Poor lighting conditions often result in a reduced dynamic range, making it difficult to distinguish between different areas of the scene, especially in shadows or highlights.

Low contrast occurs when the intensity values of an image are too similar across regions, causing the image to look flat or washed out. This can happen in foggy or dim environments or when capturing scenes with uniformly colored surfaces. In such cases, important details—like edges, textures, or object boundaries—may become indistinct, complicating tasks such as segmentation, recognition, or enhancement.

On the other hand, harsh lighting—such as direct sunlight or strong artificial lights—can produce high contrast, leading to overly bright highlights and very dark shadows. This can introduce glare, reflection, or strong shadows that obscure features or distort the true appearance of objects in the scene.

Situations where backlighting dominates the scene, such as capturing a subject in front of a window or sunset, can also create silhouette effects, where the foreground becomes dark and lacks detail. Similarly, indoor environments with inconsistent light sources or color temperatures can lead to uneven exposure and color imbalance.

In all these cases, lighting and contrast directly influence how well the image preserves structural and visual information. Managing these factors—through controlled lighting, exposure settings, or post-processing techniques—is essential for achieving high-quality images suitable for both human interpretation and automated analysis.

4.3   Cumulative Histogram

The cumulative histogram [4] is a variant of the standard histogram that provides additional information useful for performing point-wise operations on images, where each pixel is modified individually based on its intensity value. One of the primary applications of the cumulative histogram is in histogram equalization, a technique used to enhance image contrast. Unlike the standard histogram, which shows the frequency of each individual intensity level, the cumulative histogram $H(i)$ represents the total number of pixels with intensity values less than or equal to iii. In other words, $H(i)$ is the sum of all histogram values from intensity level 0 up to i. This cumulative representation allows for the calculation of normalized intensity mappings that redistribute pixel values across the available dynamic range, resulting in a more uniform histogram and improved visual quality. Because of this capability, the cumulative histogram plays a crucial role in contrast enhancement and other image preprocessing tasks. The cumulative histogram $H(i)$ can be modeled as:

$$H(i) = \sum_{j=0}^{i} h(j) \quad \text{for } 0 \leq i < K. \tag{4.2}$$

The value of the cumulative histogram $H(i)$ is obtained by summing all the values of the standard histogram $h(j)$ for intensity levels from $j = 0$ up to the specified value

i. In other words, $H(i)$ represents the total number of pixels in the image that have an intensity value less than or equal to iii. This can also be computed recursively by adding the current histogram value $h(j)$ to the cumulative value of the previous intensity level, $H(i-1)$, as expressed in Eq. 4.3. This accumulation process forms a running total that grows monotonically, reflecting the progressive distribution of pixel intensities in the image. The cumulative histogram is particularly useful for defining transformations in point-wise operations, such as those used in histogram equalization, where it helps redistribute intensity values to enhance image contrast.

$$H(i) = \begin{cases} h(0) & \text{for } i = 0 \\ H(i-1) + h(i) & \text{for } 0 \le i < K \end{cases} \tag{4.3}$$

By definition, the cumulative histogram is a monotonically increasing function, meaning that its values either stay the same or increase as the intensity level i progresses. This property arises because the cumulative histogram $H(i)$ represents the total count of pixels with intensity values less than or equal to i, which naturally accumulates as higher intensity levels are included. The function reaches its maximum value at the highest intensity level, which corresponds to the total number of pixels in the image—calculated as $M \times N$, where M and N are the image's height and width, respectively. This relationship is formally expressed in Eq. 4.4 and ensures that the cumulative histogram fully encapsulates the image's intensity distribution, making it a key tool for tasks like histogram equalization and contrast adjustment.

$$H(K-1) = \sum_{j=0}^{k-1} h(j) = M \times N. \tag{4.4}$$

Figure 4.7 illustrates a practical example of a cumulative histogram and its relationship to the original image and its intensity distribution. In Fig. 4.7a, the original grayscale image is presented, displaying the raw pixel data as captured. Figure 4.7b shows the corresponding standard histogram, which depicts the frequency of each intensity value in the image, highlighting how pixel values are distributed across the available range. Finally, Fig. 4.7c displays the cumulative histogram derived from the original histogram. This cumulative plot reveals how intensity values accumulate, with each point indicating the total number of pixels with intensity less than or equal to a given level. Together, these visualizations help illustrate how the cumulative histogram builds upon the standard histogram and provides valuable insight for image processing tasks such as contrast enhancement and histogram equalization.

The cumulative histogram is a powerful analytical tool used in various domains beyond image processing, where it originally gained prominence for tasks

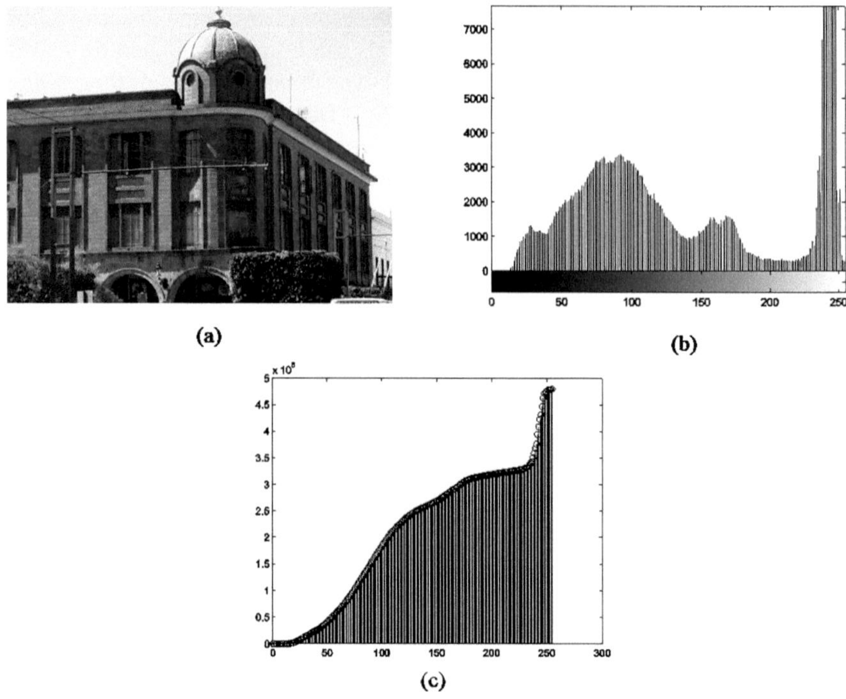

(a)

(b)

(c)

Fig. 4.7   **a** Original image, **b** histogram of (**a**), and **c** the cumulative histogram of (**a**)

like contrast enhancement and histogram equalization. In statistics and data analysis, cumulative histograms—often visualized as cumulative distribution functions (CDFs)—are used to understand the aggregated frequency of values up to a certain threshold. This helps in identifying percentile ranks, medians, or thresholds in large datasets, making it valuable in fields like economics and public health for analyzing income distributions, disease incidence rates, or population trends.

In finance, cumulative histograms are employed to study the distribution of asset returns, trading volumes, or risk levels over time. They assist in identifying the proportion of outcomes that fall below certain financial thresholds, supporting risk assessment and portfolio optimization.

In meteorology and environmental science, cumulative histograms are used to analyze temperature distributions, rainfall amounts, and pollutant concentrations. For example, they help determine how often certain temperature ranges occur or how frequently air quality measurements exceed safety limits.

In engineering and quality control, cumulative histograms are part of process monitoring systems, helping to assess how measurements of manufactured parts compare to tolerances. They enable quick identification of deviations from expected behavior and are useful in Six Sigma and other quality assurance methodologies.

In bioinformatics and medicine, cumulative histograms assist in evaluating biological measurements, such as gene expression levels, reaction times, or patient outcomes. They are crucial for determining clinical thresholds, treatment effectiveness, or the spread of values in biomedical datasets.

Across all these domains, the cumulative histogram serves as a valuable tool for understanding distributional trends, comparing datasets, setting thresholds, and supporting decision-making by offering a cumulative view of data that highlights how quantities build up over a defined range.

4.4 Computing Image Histograms Using Python

This section presents the Python functions used to calculate and display the histogram of an image, bridging theoretical concepts with practical implementation [5]. By applying the mathematical foundations discussed earlier—such as the definition of a histogram and its cumulative form—this section demonstrates how these ideas can be translated into real-world applications using Python's image processing libraries, such as OpenCV or Matplotlib. Through hands-on examples, it is possible to learn how to extract the distribution of pixel intensities from an image and visualize it effectively, enabling further analysis and processing based on contrast, brightness, and dynamic range. The goal is to reinforce the theoretical understanding of histograms through practical coding exercises that illustrate their relevance in digital image analysis.

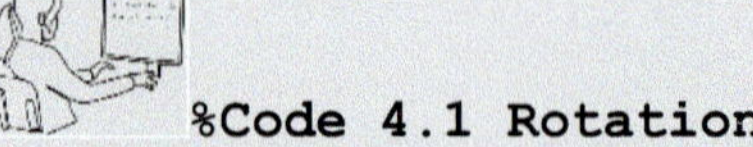

%Code 4.1 Rotation

```python
# The libraries necessary for image processing are loaded.

import cv2 as cv
import numpy as np
import matplotlib.pyplot as plt
# for image display
from google.colab.patches import cv2_imshow

img = cv.imread('imgn17.png')
cv2_imshow(img)
```

```python
# Draw the histogram line for each channel in different
colors of the image
hist = cv.calcHist([img],[0],None,[256],[0,255])
plt.plot(hist, color='blue' )
hist = cv.calcHist([img],[1],None,[256],[0,255])
plt.plot(hist, color='green' )
hist = cv.calcHist([img],[2],None,[256],[0,255])
plt.plot(hist, color='red' )

plt.xlabel('Illumination intensity')
plt.ylabel('Number of pixels')
plt.show()
```

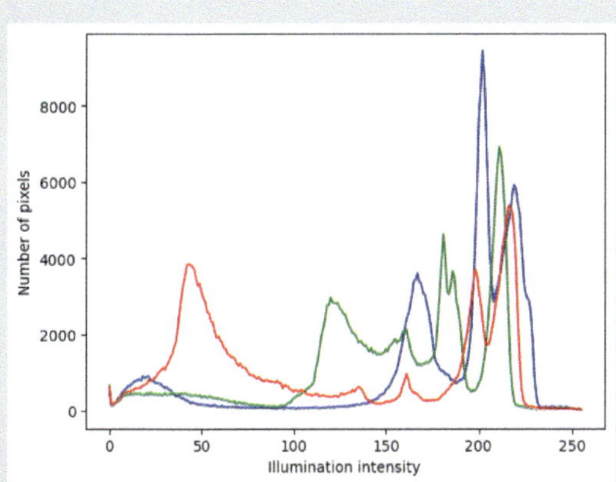

```python
#The image is transformed from color to grayscale
img = cv.cvtColor(img,cv.COLOR_BGR2GRAY)
cv2_imshow(img)
```

```python
# for display the histogram line
hist = cv.calcHist([img],None,None,[256],[0,255])
plt.plot(hist, color='gray' )

plt.xlabel('Illumination intensity')
plt.ylabel('Number of pixels')
plt.show()
```

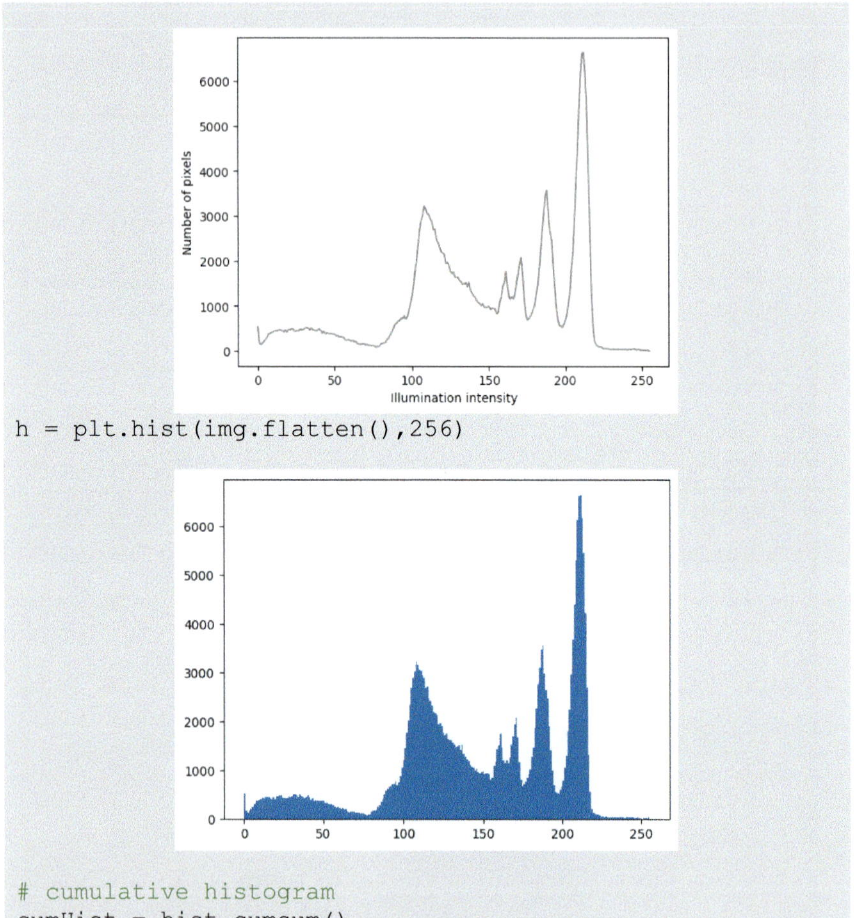


```python
h = plt.hist(img.flatten(),256)
```

```python
# cumulative histogram
cumHist = hist.cumsum()
plt.plot(cumHist)
```

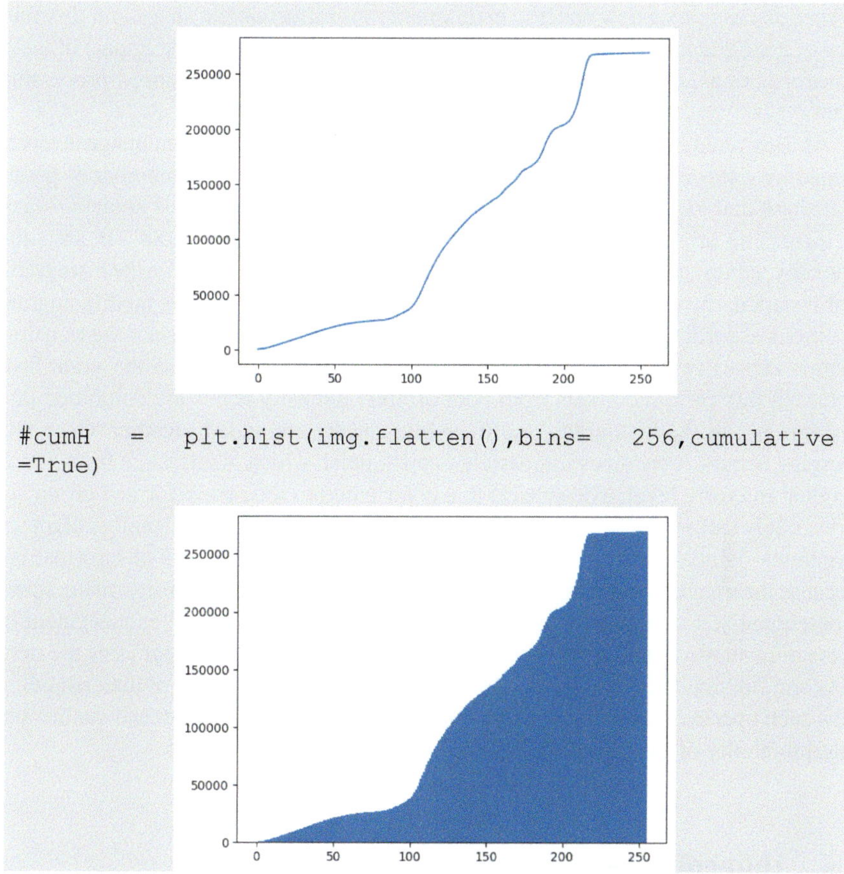

```
#cumH    =    plt.hist(img.flatten(),bins=    256,cumulative
=True)
```

4.5 Image Histograms and Point Processing Techniques

The effects of pixel operations on an image are clearly reflected in its histogram, making the histogram a valuable tool for understanding and evaluating these transformations. For example, when image illumination is increased—typically by adding a constant value to all pixel intensities—the entire histogram shifts to the right, indicating brighter pixel values. Similarly, when contrast is enhanced, the histogram expands across a wider range of intensity values, highlighting increased differentiation between dark and bright regions. In the case of image complementation or inversion, where each pixel intensity is replaced by its opposite (e.g., $255 - a$ in an 8-bit image), the histogram is mirrored with respect to the central value of the intensity range. While these effects might appear straightforward or even obvious, a deeper discussion of how specific pixel operations influence the shape and distribution of histograms can provide valuable insight into the behavior and impact of these transformations, especially when preparing images for more advanced processing steps.

As illustrated in Fig. 4.7, each homogeneous intensity region in an image is represented by a specific point i in the histogram, corresponding to the number of pixels that share that exact intensity value. When a pixel-wise operation is applied—such as increasing brightness, decreasing contrast, or applying a threshold—it alters the intensity values of pixels, which directly affects their placement within the histogram. For instance, increasing all pixel intensities by a fixed amount shifts their histogram positions accordingly. However, a more complex situation arises when the transformation causes two or more distinct intensity values to converge to the same final value. In this case, the pixels from both original intensity levels are combined into a single bin in the histogram, and their frequencies are added together. Once this merging occurs, it becomes impossible to distinguish which pixels came from which original intensity level. The subsets that contributed to the newly formed group are irreversibly lost as separate entities. This phenomenon represents a clear loss of image dynamics—since fewer intensity levels are now in use—and a loss of information, because the image can no longer retain or reproduce its original intensity distinctions. Consequently, while point-wise operations are powerful for image enhancement or correction, they must be applied carefully to avoid unintentionally degrading the richness and fidelity of the image data. This highlights the importance of understanding how such operations interact with the histogram and influence the overall quality and interpretability of the processed image.

4.6   Auto-enhancement of Image Contrast

The goal of automatic contrast adaptation is to adjust an image's pixel intensity values so that they span the entire available range, thereby enhancing visual clarity and making better use of the dynamic range. This process involves identifying the darkest and brightest pixels in the image and mapping them to the minimum and maximum intensity values allowed—for example, 0 and 255 in an 8-bit image. All other pixel values falling between these extremes are then linearly interpolated to fit within this expanded range. As a result, the intensity values are redistributed more evenly, increasing contrast and revealing details that may have been hidden in underexposed or overexposed regions. This transformation is particularly useful in images where the original contrast is low due to poor lighting or sensor limitations, as it enhances visibility without introducing new artifacts or altering the spatial structure of the image.

In the process of automatic contrast adjustment, it is assumed that P_{low} and P_{high} represent the actual minimum and maximum intensity values found in an image III, which originally contains pixel values within the interval $[P_{\text{min}}, P_{\text{max}}]$. To enhance contrast and utilize the full available intensity scale, the algorithm maps P_{low} to the

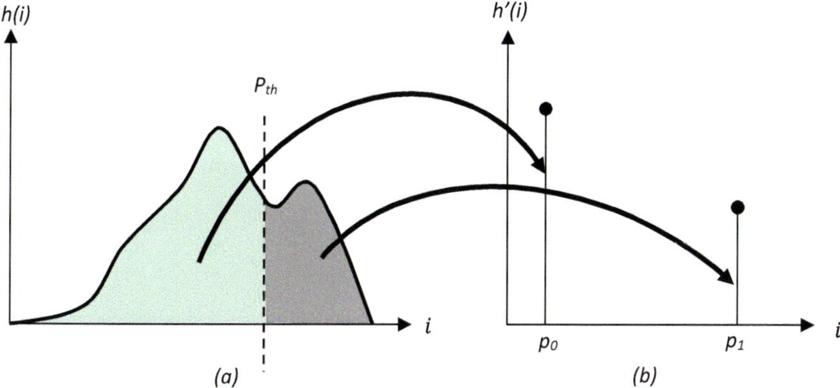

Fig. 4.8 Effect of the binarization operation on the histogram. The threshold value is P_{th}. **a** Original histogram and **b** histogram resulting from the operation, concentrating their values at two points, p_0 and p_1

minimum allowable value (typically 0 in an 8-bit grayscale image) and P_{high} to the maximum allowable value (usually 255). All intermediate pixel values are linearly rescaled based on these two reference points. This operation effectively stretches the histogram across the full dynamic range, improving contrast by increasing the difference between light and dark regions. As illustrated in Fig. 4.8, this contrast enhancement is achieved by applying a scaling factor defined in Eq. 4.5 that redistributes the pixel intensities proportionally, ensuring that the full range of tonal values is used to represent image details more vividly and accurately (Fig. 4.9).

$$\frac{P_{\max} - P_{\min}}{P_{high} - P_{low}}. \tag{4.5}$$

Hence, the most basic form of contrast adjustment can be expressed as follows:

$$f_{ac} = (p - p_{low}) \cdot \left(\frac{P_{\max} - P_{\min}}{P_{high} - P_{low}} \right). \tag{4.6}$$

For an 8-bit grayscale image, the intensity values are defined within the fixed range of $p_{\min} = 0$ and $p_{\max} = 255$. This predefined range allows the contrast adaptation function to be simplified, since the minimum and maximum allowable values are already known and fixed. As a result, the transformation equation that adjusts pixel intensities to span the full dynamic range becomes more straightforward, eliminating the need to reference arbitrary bounds. This simplification is clearly shown in Eq. 4.7, where the formula directly maps the lowest and highest intensity values found in the image to 0 and 255, respectively, while linearly scaling the intermediate values. This streamlined approach facilitates efficient implementation and ensures consistent contrast enhancement for standard 8-bit images (Fig. 4.10).

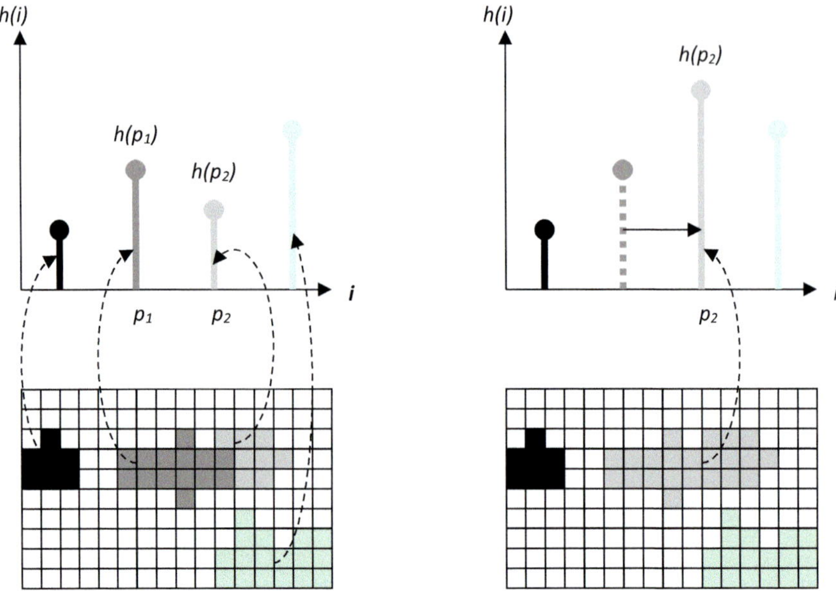

Fig. 4.9 Impact of the binarization process on the histogram using a threshold value P_{th}: **a** the original histogram, and **b** the transformed histogram, where pixel intensities are reduced to two distinct values, p_0 and p_1

$$f_{\text{ac}} = (p - p_{\text{low}}) \cdot \left(\frac{255}{P_{\text{high}} - P_{\text{low}}} \right). \tag{4.7}$$

The interval $[P_{\text{min}}, P_{\text{max}}]$ used in contrast adaptation does not necessarily have to correspond to the full allowable intensity range of an image; rather, it can represent any subinterval within the permissible limits of image representation. This flexibility allows for targeted contrast adjustments based on specific intensity regions of interest. When the selected interval is narrower than the actual intensity range of the image— meaning $p_{\text{min}} > p_{\text{low}}$ and $p_{\text{max}} < p_{\text{high}}$—applying the contrast adaptation method will

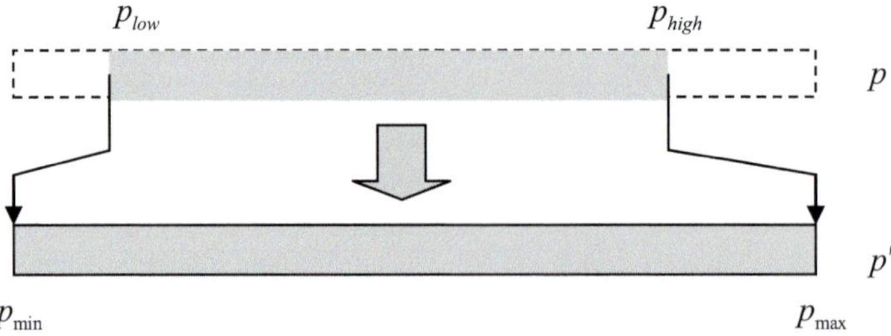

Fig. 4.10 Auto-contrast operation based on Eq. 4.5, where the pixel value p is linearly transformed from the original range $[P_{\text{low}}, P_{\text{min}}]$ to the full intensity range $[P_{\text{min}}, P_{\text{max}}]$

result in a reduction of contrast rather than an enhancement. This happens because the pixel values are compressed into a smaller range, thereby diminishing the intensity differences across the image. On the other hand, if the selected interval spans a broader range, the operation enhances contrast by stretching the pixel values across a wider spectrum. Figure 4.11 illustrates the visual outcome of an auto-contrast operation, showing how changes to the reference interval directly influence the image's contrast and dynamic appearance.

In practical applications, the automatic contrast adjustment method defined by Eq. 4.6 may sometimes yield unsatisfactory results due to its sensitivity to extreme pixel values. Specifically, the values p_{low} and p_{high}, which represent the minimum and maximum pixel intensities in the original image, may correspond to only a very small number of pixels. These outliers, while mathematically valid, do not significantly reflect the overall intensity distribution of the image. As a result, using them to stretch the pixel values across the full dynamic range can lead to misleading or distorted contrast adjustments—either drastically altering the histogram or barely affecting it at all. To address this issue, a more robust approach is adopted by introducing fixed percentage thresholds, denoted as S_{low} and S_{high}, which define what portion of the darkest and brightest pixels should be ignored in the contrast enhancement

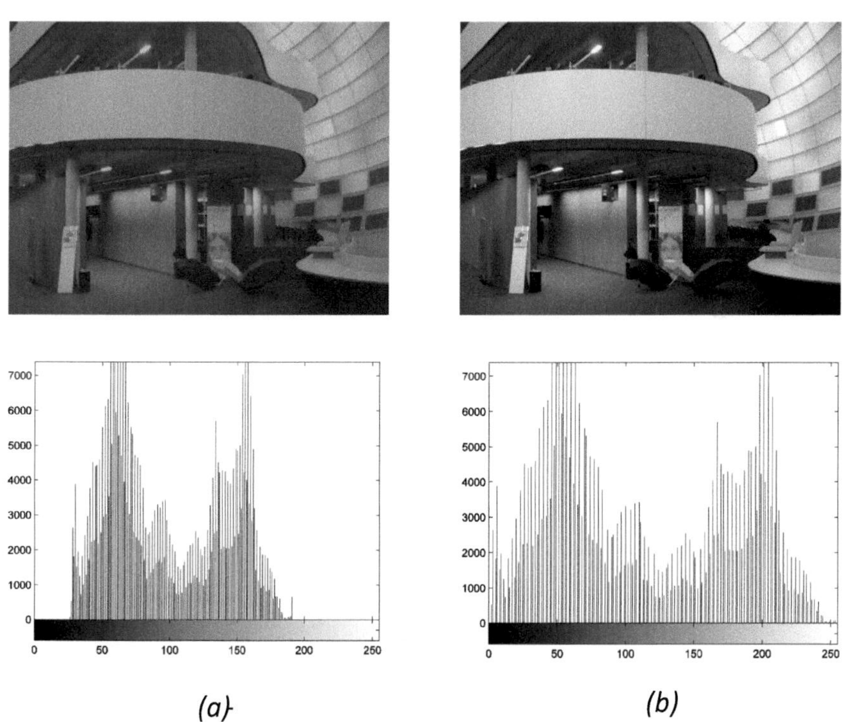

Fig. 4.11 Impact of the auto-contrast operation: **a** the original image along with its corresponding histogram, and **b** the image after applying auto-contrast and its updated histogram

process. Instead of directly using the extreme values, the algorithm determines two new boundaries, a_{low} and a_{high}, that mark the beginning and end of the "significant" intensity range. The value a_{low} is calculated as the smallest intensity level such that the cumulative number of pixels with intensities less than or equal to it is at least S_{low} of the total pixel count. Similarly, a_{high} is identified as the highest intensity level such that the cumulative number of pixels with intensities greater than or equal to it is no more than S_{high} of the total. This process ensures that the contrast adjustment reflects the majority of the image content while ignoring insignificant outliers. Figure 4.12 visually demonstrates how these thresholds are determined, and Eqs. 4.8 and 4.9 formally define the computation of a_{low} and a_{high} using the cumulative histogram $H(i)$ of the image I. This refined method offers a more stable and meaningful contrast enhancement by grounding the transformation in a statistically representative subset of pixel values.

$$a_{\text{low}} = \min\{i|H(i) \geq M \cdot N \cdot S_{\text{low}}\}, \tag{4.8}$$

$$a_{\text{high}} = \min\{i|H(i) \leq M \cdot N \cdot (1 - S_{\text{high}})\}. \tag{4.9}$$

In the context of automatic contrast enhancement, $M \cdot N$ represents the total number of pixels in the image I. To ensure that the contrast adjustment focuses only on the most relevant portions of the image, all intensity values that fall outside the interval $[a_{\text{low}}, a_{\text{high}}]$ are excluded from the enhancement process. This means that only the pixel values within this interval—those deemed statistically significant based on the

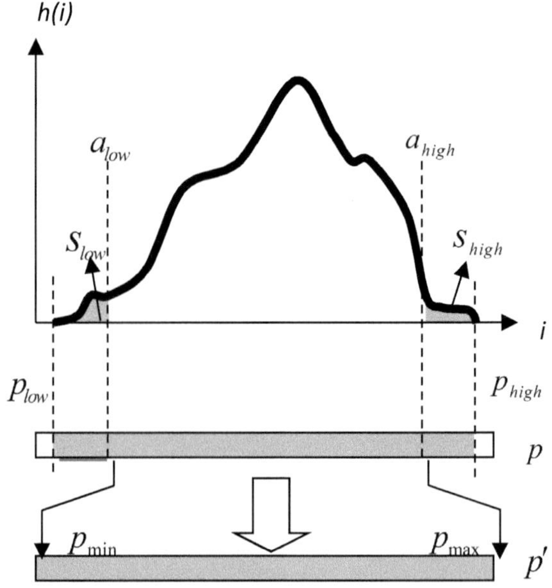

Fig. 4.12 Auto-contrast operation uses percentage thresholds to define boundaries that focus only on the significant portion of the intensity distribution during adjustment. Pixel values below a_{low} and above a_{high} (within the 0–255 range) are excluded from the adjustment, allowing the resulting image to exhibit enhanced contrast based on the most relevant intensity values

cumulative histogram—are considered for contrast adaptation. These selected values are then linearly scaled to fill the entire permissible intensity range $[p_{\min}, p_{\max}]$, such as [0, 255] in an 8-bit grayscale image. This transformation effectively stretches the meaningful part of the intensity distribution across the full dynamic range, improving contrast while ignoring extreme outliers that could distort the result. The actual pixel-level transformation that carries out this auto-contrast adjustment is mathematically defined by Eq. 4.10, which formalizes the mapping from the original intensity values to their enhanced counterparts based on the computed boundaries.

$$f_{\max} = \begin{cases} p_{\min} & \text{if } p \leq a_{\text{low}} \\ (p - a_{\text{low}}) & \text{if } a_{\text{low}} < p < a_{\text{high}} \\ p_{\max} & \text{if } p \geq a_{\text{high}} \end{cases} . \tag{4.10}$$

In practical implementations of automatic contrast adjustment, the values of S_{low} and S_{high} are typically set to the same fixed value, denoted as S, which usually falls within the interval [0.5, 1.5]. This small percentage is sufficient to exclude extreme outlier pixels that do not significantly contribute to the overall visual content of the image, thereby ensuring a more robust and representative contrast enhancement. A well-known commercial example of this technique can be found in Adobe Photoshop, where the default value of S is set to 0.5% during the automatic contrast adjustment process. This means that the lowest and highest 0.5% of pixel intensities are disregarded, and the remaining intensity values are scaled to occupy the full available dynamic range. This approach improves image visibility and detail without being overly influenced by rare, extreme intensity values that could otherwise distort the adjustment.

References

1. Burger, W., & Burge, M. J. (2022). Histograms and image statistics. *Digital image processing: An algorithmic introduction* (pp. 29–48). Springer International Publishing.
2. Williams, M. B., Krupinski, E. A., Strauss, K. J., Breeden III, W. K., Rzeszotarski, M. S., Applegate, K., Wyatt, M., Bjork, S., & Seibert, J. A. (2007). Digital radiography image quality: image acquisition. *Journal of the American College of Radiology, 4*(6), 371–388.
3. Kotkar, V. A., & Gharde, S. S. (2013). Review of various image contrast enhancement techniques. *International Journal of Innovative Research in Science, Engineering and Technology, 2*(7).
4. Garg, P., & Jain, T. (2017). A comparative study on histogram equalization and cumulative histogram equalization. *International Journal of New Technology and Research, 3*(9), Article 263242.
5. Chityala, R., & Pudipeddi, S. (2020). *Image processing and acquisition using Python.* Chapman and Hall/CRC.

Chapter 5
Spatial Filters

In previous chapters, we explored pixel-based operations, where the updated value of a pixel is determined exclusively by its original value and is reassigned to the exact location. While these operations can produce a variety of visual effects, they fall short in specific scenarios, such as image smoothing illustrated in Fig. 5.1, or edge detection, where more complex manipulations are required. In such cases, filters come into play. Filtering operations compute each new pixel value not only from the target pixel itself but also from its surrounding neighbors within a defined area. This approach enables more advanced image transformations while preserving the spatial structure of the image. In other words, despite involving multiple pixels, the filter does not alter the image's geometry, maintaining a consistent correspondence between the original and processed image, like what occurs in simple pixel-by-pixel operations.

5.1 Understanding Image Filters

In digital images, certain areas exhibit sudden local changes in brightness, sharp increases or decreases in intensity from one pixel to the next. In contrast, some regions maintain a steady intensity with little to no variation [1]. The concept behind image smoothing is to reduce these abrupt transitions by averaging intensity values, effectively softening the visual appearance.

A basic method for smoothing involves replacing each pixel with the average value of its surrounding neighbors. To compute the value of a smoothed pixel, denoted as $I'(x, y)$, we consider the original pixel $I(x, y) = p_0$ along with its eight immediate neighbors p_1 through p_8. The new value is obtained by calculating the arithmetic meaning of these nine pixels:

$$I'(x, y) \leftarrow \frac{p_0 + p_1 + p_2 + p_3 + p_4 + p_5 + p_6 + p_7 + p_8}{9}.$$
(5.1)

© The Author(s), under exclusive license to Springer Nature Switzerland AG 2026

E. Cuevas et al., *Image Processing with Python*, Signals and Communication Technology, https://doi.org/10.1007/978-3-032-13285-7_5

(a)   (b)

Fig. 5.1 Pixel-based operations cannot achieve effects like smoothing. **a** Original image, **b** smoothed result

This process helps to diminish sharp variations in intensity, producing a more uniform and visually pleasant result.

Using coordinates that reference the image itself, the process described can be formulated as follows:

$$I'(x, y) \leftarrow \frac{1}{9} \cdot \begin{bmatrix} I(x-1, y-1) + I(x, y-1) + I(x+1, y-1) \\ +I(x-1, y) + I(x, y) + I(x+1, y) \\ +I(x-1, y+1) + I(x, y+1) + I(x+1, y+1) \end{bmatrix}. \tag{5.2}$$

This can be represented in a simplified, concise format:

$$I'(x, y) \leftarrow \frac{1}{9} \cdot \sum_{j=-1}^{1} \sum_{i=-1}^{1} I(x+i, y+j). \tag{5.3}$$

The computed local average illustrates the fundamental components commonly found in filtering techniques. This approach serves as a clear example of a widely used category known as *linear filters*, where output values are derived from a linear combination of neighboring pixel intensities (Fig. 5.2).

The dimensions of the neighborhood region $R(x, y)$ play a crucial role in filtering, as they define both the number and the specific arrangement of pixels from the original image that will influence the new pixel value. In the earlier example, a 3×3 smoothing filter centered at coordinate (x, y) was used. Expanding the size of this region, such as to 5×5, 9×9, or even 33×33 pixels, results in a more substantial smoothing effect, as more surrounding data points contribute to the final output.

Although the region's shape can theoretically vary, square regions are the most commonly employed. It is due not only to computational simplicity (as opposed to circular or irregular shapes requiring complex calculations) but also because squares evenly sample neighboring pixels in all directions, which is typically beneficial in image processing.

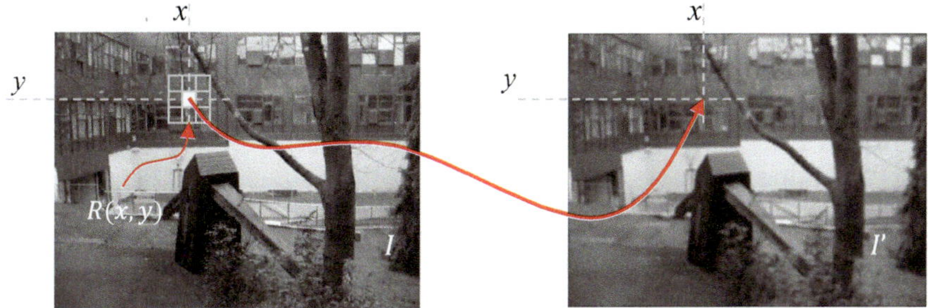

Fig. 5.2 Each output pixel $I'(x, y)$ is determined by analyzing a specific neighborhood region $R(x, y)$ centered around the target location. The coordinates of this region are aligned with the position of the pixel being processed, ensuring that the local context directly influences the resulting value

Filters can also differ in how they treat each pixel within the region. Rather than treating all neighboring pixels equally, a weighting scheme can be applied, giving more influence on those closer to the center and less to those farther away. This weighted approach improves the filter's responsiveness to local variations and structure.

Given these variations, there are many possible ways to construct a filter. However, a structured approach is essential to define, design, and implement filters effectively. Additionally, filters may follow either linear or nonlinear models, each with unique operational behaviors and effects. The key distinction lies in how the pixel values within the region are combined either through linear combinations or more complex nonlinear operations.

As introduced in this chapter, filtering involves summing the products of each pixel in the neighborhood with a corresponding coefficient from a predefined matrix. This matrix, commonly referred to as a filter, mask, kernel, or window in image processing, governs how each pixel contributes to the final result. The next chapter will explore both linear and nonlinear filters in greater detail, accompanied by real-world examples to illustrate their application.

5.2 Spatial Linear Filters

The term linear filter refers to operations in which the intensity values of pixels within a defined neighborhood are combined through a linear process to generate the new pixel value. A simple illustration of this idea appears in Eq. 5.3, where nine pixels in a 3×3 region are averaged by summing their values and multiplying the result by the factor (1/9). Using the same principle, many other filters can be designed, each producing different effects on an image, depending solely on how the pixel values in the neighborhood are weighed and combined [2].

5.2.1　The Filter Matrix

To completely define a linear filter, two aspects must be specified: the geometry of the processing region (its size and shape) and the set of coefficients that dictate how pixel values are linearly combined. These coefficients, often called weights, are organized into what is known as the filter matrix or mask.

The dimensions of this matrix establish the size of the neighborhood over which the filter operates. At the same time, its numerical entries determine the relative importance assigned to each pixel in the region. When applied, each pixel in the neighborhood is multiplied by its corresponding coefficient, and the results are summed to produce the new pixel value.

As an example, the smoothing filter described in Eq. 5.3 can be fully expressed through the following filter matrix:

$$
H(i,j) = \begin{bmatrix} 1/9 & 1/9 & 1/9 \\ 1/9 & 1/9 & 1/9 \\ 1/9 & 1/9 & 1/9 \end{bmatrix} = \frac{1}{9} \begin{bmatrix} 1 & 1 & 1 \\ 1 & 1 & 1 \\ 1 & 1 & 1 \end{bmatrix}. \tag{5.4}
$$

Each of the nine pixels included in the 3×3 mask contributes equally to the calculation of the new pixel value, with each one providing one-ninth of the total.

From a mathematical perspective, the filter matrix $H(I,j)$ can be described as a two-dimensional discrete real function, in a similar way to how an image itself is represented. The coordinate system of the matrix is centered at its origin, located in the middle of the grid, which means its indices can take both positive and negative values (see Fig. 5.3). Any coefficients that fall outside the defined coordinates of the matrix are assumed to be zero [3].

Fig. 5.3 Filter matrix (grid) and its coordinate system

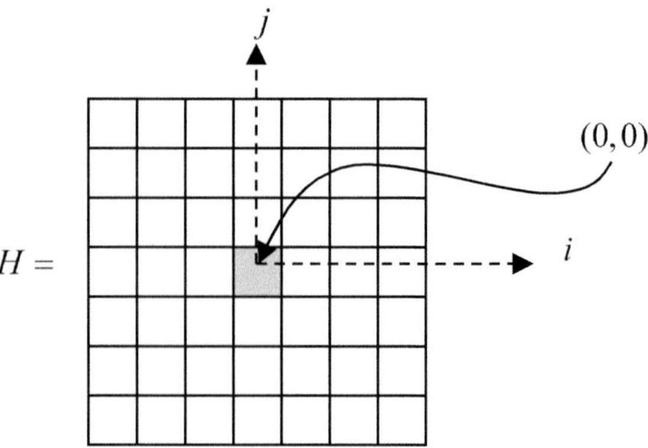

5.2.2 *Filter Operation*

The effect of applying a filter to an image is entirely determined by the coefficients of its matrix $H(i, j)$. The procedure, illustrated in Fig. 5.4, can be summarized in the following steps:

1. The filter matrix $H(i, j)$ is aligned with the input image I placing its center at pixel $H(0, 0)$ on top of the pixel of interest $I(x, y)$.
2. Each pixel of the image inside the neighborhood region $R(x, y)$ defined by the size of the matrix, is multiplied by the corresponding coefficient in $H(i, j)$. The results of these multiplications are then added together.
3. The final sum is assigned to the corresponding pixel $I'(x, y)$ in the output image.

This process can be expressed mathematically as:

$$I'(x, y) = \sum_{(i,j)\in R(x,y)} I(x + i, y + j) \cdot H(i, j). \tag{5.5}$$

For a typical 3×3 filter, this becomes:

$$I'(x, y) = \sum_{j=-1}^{1} \sum_{i=-1}^{1} I(x + i, y + j) \cdot H(i, j). \tag{5.6}$$

Applied to all valid coordinates (x, y) of the image. It is important to note, however, that this formulation cannot be applied directly to pixels located at the borders of the image. At those positions, the filter extends beyond the image boundaries, leading to undefined results. This boundary problem will be revisited, along with possible solutions, in Sect. 5.5.2.

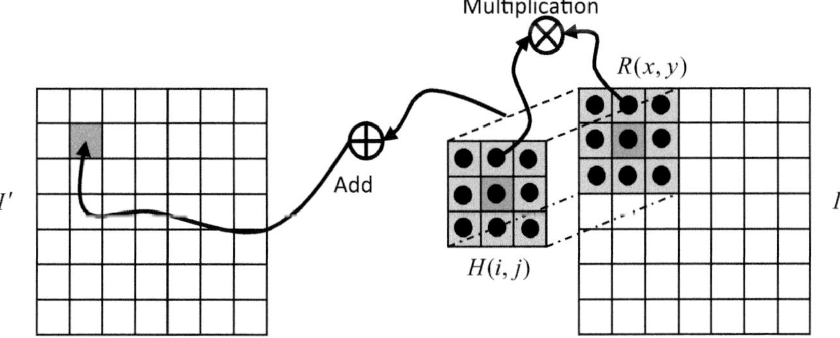

Fig. 5.4 Filter matrix is centered on the pixel (x, y) of the original imagen I. Each coefficient $H(i, j)$ is multiplied by the corresponding image pixel, and the nine products are summed. The resulting value is then assigned to the pixel at (x, y) in the output image I'

5.3 Implementing Filter Operations in Python

Once the concept of filtering and its implications at the image boundaries are understood, the next step is to see how these operations can be implemented programmatically using Python [4].

In spatial filtering, the procedure is carried out through a convolution process. It involves combining the pixel of interest (x, y), with its neighboring pixels and the corresponding coefficients of the filter matrix. The general idea of this mechanism is illustrated in Fig. 5.4. A key aspect of programming spatial filters is handling the alignment between the filter's coordinates and those of the image. Essentially, the filter is placed on top of the image so that its center coincides with a target pixel. The filter is then systematically shifted across each row from left to right, and then down column by column, computing the weighted sum at each position. This process is depicted in Fig. 5.5. For a filter of dimensions $m \times n$, the sizes can be expressed as: $m = 2a+1$ and $n = 2b+1$ where a and b are non-negative integers. This formulation suggests that filter matrices are typically of odd size, with the smallest commonly used being a 3×3. While it is not strictly necessary to use odd dimensions, doing so greatly simplifies the process. With an odd-sized filter, there is a single, well-defined center, which ensures that the operation remains symmetric and computationally straightforward.

The pixel value produced at location (x, y) after applying the filter is determined by combining the intensity values of the image with the corresponding coefficients within the neighborhood region $R(x, y)$. Based on the coefficients of the filter and the image intensities illustrated in Fig. 5.6, the resulting pixel value can be expressed as:

$$I'(x, y) = I(x - 1, y - 1) \cdot H(-1, -1) + I(x - 1, y)$$
$$\cdot H(-1, 0) + I(x - 1, y + 1) \cdot H(-1, 1)$$

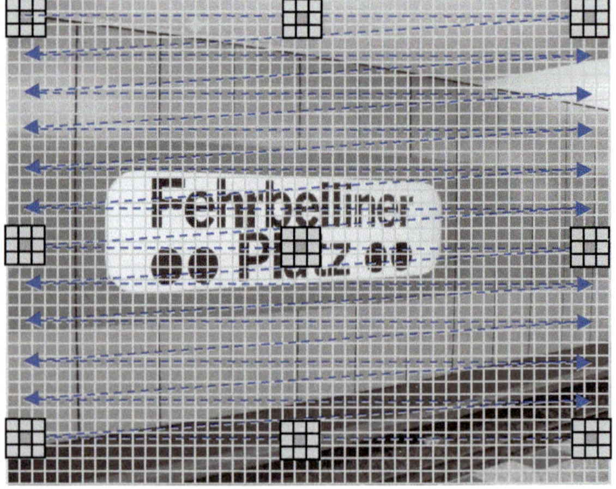

Fig. 5.5 Apply the spatial filter to an image

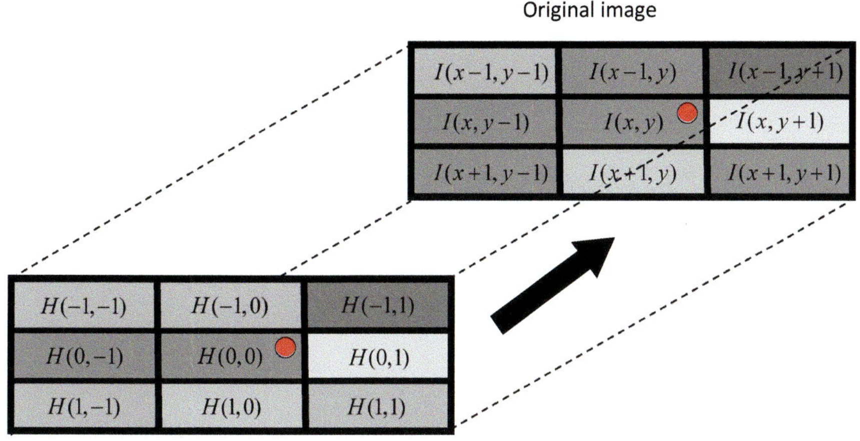

Fig. 5.6 Spatial filtering process

$$+ I(x, y - 1) \cdot H(0, -1) + I(x, y)$$
$$\cdot H(0, 0) + I(x, y + 1) \cdot H(0, 1)$$
$$+ I(x + 1, y - 1) \cdot H(1, -1) + I(x + 1, y)$$
$$\cdot H(1, 0) + I(x + 1, y + 1) \cdot H(1, 1). \tag{5.7}$$

A crucial consideration when implementing filters is the impact of image boundaries on the result. When the filter's coefficient matrix is centered over every pixel of the image, some positions of the matrix H inevitably extend beyond the image domain, leaving them without corresponding pixel values in the image I. This boundary condition, along with strategies for its treatment, will be examined in a subsequent section. A graphical representation of the problem is provided in Fig. 5.7.

A simple approach to deal with the boundary problem is to limit the filtering process to a smaller portion of the image. Instead of moving the filter center across the entire range of coordinates $1 \ldots N$ and $1 \ldots M$, the traversal is restricted to $2 \ldots N - 1$ and $2 \ldots M - 1$. y doing so, every coefficient in the filter mask aligns with a valid pixel, avoiding undefined computations.

This solution, however, comes with a compromise: the border regions of the image are excluded from processing. To overcome this limitation, two widely used alternatives can be applied:

1. Reduced output size: The resulting image is smaller than the original, with dimensions $M - 2 \times N - 2$, since the border pixels are excluded from processing.
2. Preservation of image size: The original image is first copied into the output image. The filtering operation is then applied only to the interior pixels (from 2 to $N - 1$ and from 2 to $M - 1$), while the border pixels remain unchanged. This results in an output image that retains the exact dimensions of the input.

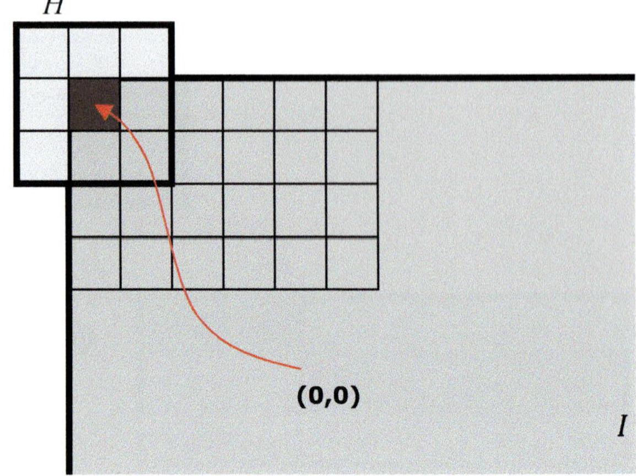

Fig. 5.7 One of the main difficulties in convolution-based filtering arises at the borders of the image. When the filter is positioned at pixel (0, 0), parts of the coefficient matrix H extend outside the image boundaries, resulting in coefficients without corresponding pixel values

5.4   Types of Linear Filters

The values within its coefficient matrix completely determine the behavior of a linear filter. Since this matrix can take on countless combinations of numbers, the range of possible linear filters is essentially unlimited. It leads to key questions: What visual effect does a given filter produce? Moreover, which filter is best suited for a specific purpose? In practical applications, linear filters are generally grouped into two main categories: smooth filters and difference filters [5].

5.4.1   Smooth Filters

The filters described earlier inherently produce a smoothing effect on images. In fact, any linear filter whose coefficient matrix is made up exclusively of positive values will blur or soften the image to some degree. The output can be thought of as a scaled average of the pixel values within the region $R(x, y)$ that the filter covers.

The "Box" Filter

One of the simplest and earliest examples of a smoothing filter is the Box filter. It is named for its block-like shape, which resembles a box when visualized in three dimensions. In this filter, all coefficients carry the same value, meaning every pixel within the region of influence is weighted equally. Because of this uniformity, the Box filter smooths images by averaging surrounding pixels. However, its abrupt edges and frequency response make it a fundamental tool—easy to implement but not ideal when higher-quality results are desired. The filter shown in Fig. 5.8a has the appearance of a box—hence the origin of its name. Because all its coefficients

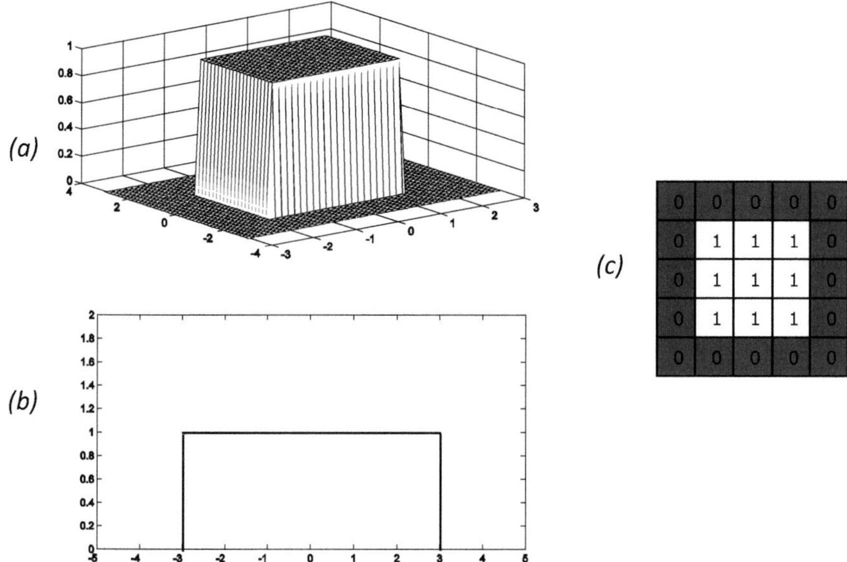

Fig. 5.8 "Box" smoothing filter. **a** Three-dimensional function, **b** two-dimensional function, and **c** mask implementing the filter

are identical, it uniformly influences every pixel of the image, giving equal weight to all positions within the filter window.

Intuitively, it would seem that a filter is less convenient in affecting all pixels of the image equally and better in giving greater weight to the central point (the main pixel) of the filter and considering the neighboring pixels in a smaller proportion. A desirable feature of smoothing filters is also the fact that they are isotropic, i.e., the effect they have on the image is invariant to rotation, which is not the case for the "Box" filter due to its quadratic form.

The Gaussian Filter

The Gaussian filter is based on the two-dimensional discrete Gaussian function, expressed mathematically as:

$$G_\sigma(r) = e^{-\frac{r^2}{2\sigma^2}} \quad \text{or} \quad G_\sigma(x, y) = e^{-\frac{x^2+y^2}{2\sigma^2}} \tag{5.8}$$

where the parameter σ represents the standard deviation of the function and effectively controls the "spread" or radius of influence of the filter, this is illustrated in Fig. 5.9.

The center of the Gaussian mask carries the most significant coefficient, meaning it contributes most strongly to the output pixel value. As the coefficients move farther from the center, their values decrease, gradually reducing their influence.

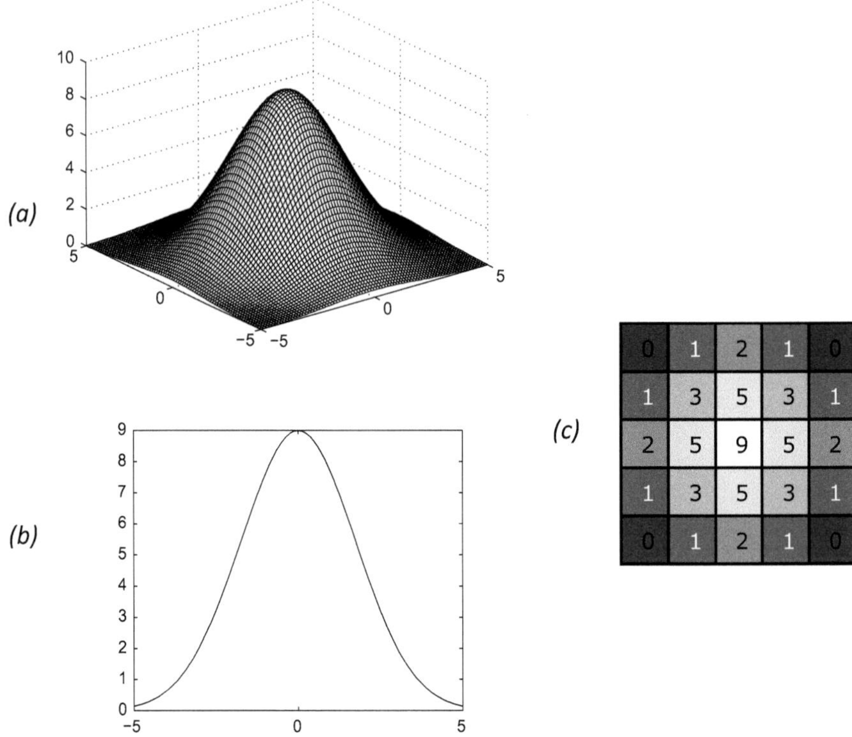

Fig. 5.9 Gaussian filter for smoothing which shows **a** the three-dimensional representation, **b** its two-dimensional projection, and **c** the mask used to implement the filter

A key characteristic of the Gaussian filter is its isotropy—it behaves the same way in all directions. It is a significant improvement over the simple Box filter, which tends to blur edges indiscriminately. In contrast, the Gaussian filter smooths areas of relatively uniform intensity while better preserving edges, resulting in a more natural-looking image.

5.4.2 Difference Filters

Unlike smoothing filters, which contain only positive coefficients, difference filters include negative values in their coefficient matrix. The presence of negative coefficients means that their operation can be interpreted as taking the difference between two sums: The sum of the weighted pixel values corresponding to the positive coefficients, and the sum of the weighted pixel values corresponding to the negative coefficients.

Mathematically, this can be expressed as:

$$I'(x, y) = \sum_{(i,j)\in R^+} I(x+i, y+j) \cdot |H(i,j)| - \sum_{(i,j)\in R^-} I(x+i, y+j) \cdot |H(i,j)| \quad (5.9)$$

where R^+ represents the region of the filter where $H(i, j) > 0$ and R^- represents the region where $H(i, j) < 0$. A classic example of this type of filter is the Laplace filter (shown in Fig. 5.10). In a 5×5 in Fig. 5.10, Laplace mask, the center pixel has a strong positive coefficient (for instance, 16), while the surrounding 12 positions have negative coefficients (such as -1 or -2). The remaining positions are zero and do not affect the computation.

This configuration effectively measures the contrast between the central pixel and its neighbors. The resulting function is commonly referred to as the "Mexican hat" function, and is mathematically defined as:

$$M_\sigma(r) = \frac{1}{\sqrt{2\pi}\sigma^3}\left(1 - \frac{r^2}{\sigma^2}\right)e^{-\frac{r^2}{2\sigma^2}}$$

or, in two-dimensional form:

$$M_\sigma(x, y) = \frac{1}{\sqrt{2\pi}\sigma^3}\left(1 - \frac{x^2 + y^2}{\sigma^2}\right)e^{-\frac{x^2+y^2}{2\sigma^2}}. \quad (5.10)$$

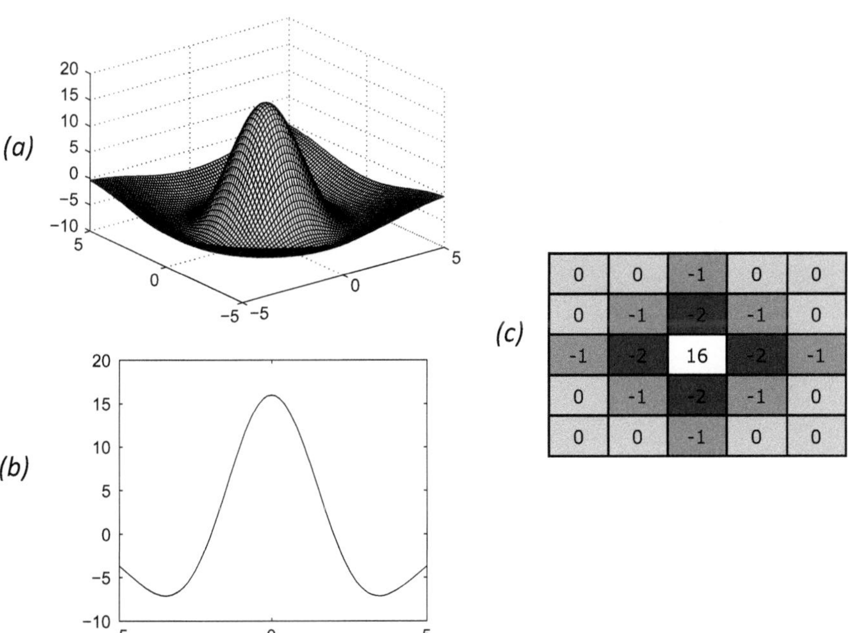

Fig. 5.10 An example: the Laplace filter, also known as the Mexican hat filter. The figure shows **a** three-dimensional representation, **b** two-dimensional function, and **c** mask used to implement it

While filters with only positive coefficients produce a smoothing effect, difference filters have the opposite effect, enhancing intensity changes and making edges and transitions more pronounced. For this reason, they are widely used in edge detection and image sharpening applications.

Linear filters in Python

#Code 3.1

```python
#The libraries necessary for image processing are loaded
import cv2 as cv
import numpy as np
import matplotlib.pyplot as plt
from google.colab.patches import cv2_imshow

# for image display
img = cv.imread('Figure0512.png')
cv2_imshow(img)
```

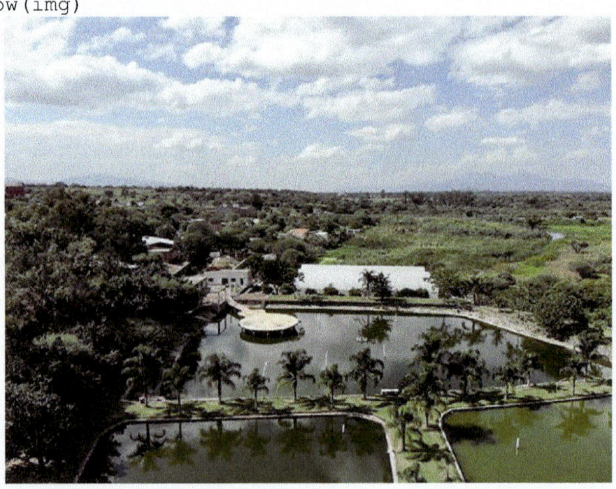

```python
# BOX filter

# Box filter with blur on RGB image
blur = cv.blur(img,(9,9))
cv2_imshow(blur)
```

```python
# Box filter with blur on grayscale image
imgGray = cv.cvtColor(img,cv.COLOR_BGR2GRAY)
blurGray = cv.blur(imgGray,(9,9))
cv2_imshow(blurGray)
```

```python
# Box filter with boxfilter on RGB image
blur2 = cv.boxFilter(img,-1,(9,9))
cv2_imshow(blur2)
```

```
# Box filter with boxfilter on grayscale image
blurGray2 = cv.boxFilter(imgGray,-1,(9,9))
cv2_imshow(blurGray2)
```

```
# Box filter with filter2D on RGB image
kernel = np.ones((9,9),np.float32)/81
blur3 = cv.filter2D(img,-1,kernel)
cv2_imshow(blur3)
```

```
# Box filter with filter2D on RGB image
blurGray3 = cv.filter2D(imgGray,-1,kernel)
cv2_imshow(blurGray3)
```

Gaussian filter

```
# Gaussian filter with GaussianBlur on RGB image
gauss = cv.GaussianBlur(img,(5,5),0)
cv2_imshow(gauss)
```

```python
# Gaussian filter with GaussianBlur on grayscale image
gaussGray = cv.GaussianBlur(imgGray,(5,5),0)
cv2_imshow(gaussGray)
```

```python
# Gaussian filter with filter2D on RGB image
kernelGauss = np.float32([[0,1,2,1,0], [1,3,5,3,1],
[2,5,9,5,2], [1,3,5,3,1], [0,1,2,1,0]]) / 57
gauss2 = cv.filter2D(img,-1,kernelGauss)
cv2_imshow(gauss2)
```

```python
# Gaussian filter with filter2D on grayscale image
gauss2Gray = cv.filter2D(imgGray,-1,kernelGauss)
cv2_imshow(gauss2Gray)
```

```python
# Gaussian filter with filter2D on grayscale image
#kernel = cv.getGaussianKernel(5,2,ktype=cv.CV_32F) #ktype is
optional
kernel = cv.getGaussianKernel(ksize=5,sigma=1) #Genera un
filtro 1D en lugar de 2D, se ocupa sepFilter2D para aplicar
este kernel
imgKernel = cv.sepFilter2D(imgGray,-1,kernel,kernel)
```

```
cv2_imshow(imgKernel)
```

5.5 Formal Characteristics of Linear Filters

Up to this point, filters have been introduced in a simplified manner to provide an intuitive sense of their purpose and construction. While this approach helps gain a quick and practical understanding, there are cases where a more profound insight is required. Specifically, understanding the mathematical foundations behind filtering operations enables greater flexibility in both designing and analyzing filters.

At the heart of linear filter operations lies the process of convolution. This mathematical tool has been briefly mentioned in earlier sections, but its formal properties are essential for explaining why filters behave as they do. A more detailed discussion of convolution and its principal characteristics will therefore follow in the next section.

5.5.1 Linear Convolution and Correlation

The use of linear filters in image processing is not a recent development but rather an application of a well-established mathematical operation: linear convolution. Convolution provides a method for combining two functions—whether continuous or discrete—to produce a single output function.

For two discrete two-dimensional functions, I (image) and H (the filter), convolution is formally defined in equation:

$$I'(x, y) = \sum_{i=-\infty}^{\infty} \sum_{j=-\infty}^{\infty} I(x - i, y - j) \cdot H(i, j). \tag{5.11}$$

Or in compact form:

$$I' = I * H. \tag{5.12}$$

The expression of Eq. 5.11 resembles the earlier Eq. 5.6, with two key differences: (1) Summation limits: The filter $H(i, j)$ defines a region $R(x, y)$ Coefficients outside this region are considered zero and do not affect the computation. For this reason, the summation can extend beyond $R(x, y)$ without altering the outcome. (2) Coordinate inversion: In convolution, the coordinates are shifted to $I(x - 1, y - j)$, which differs from Eq. 5.6. Suppose the formulation is rearranged as shown in equation:

$$I(x, y) = \sum_{(i,j)\in R} I(x - i, y - j) \cdot H(i, j) = \sum_{(i,j)\in R} I(x + i, y + j) \cdot H(-i, -j). \tag{5.13}$$

In that case, it becomes clear that convolution is essentially the same operation as before, except the filter matrix $H(i, j)$ is inverted in both axes, yielding $H(-i, -j)$. Geometrically, this inversion corresponds to a 180° rotation of the filter mask, as illustrated in Fig. 5.11.

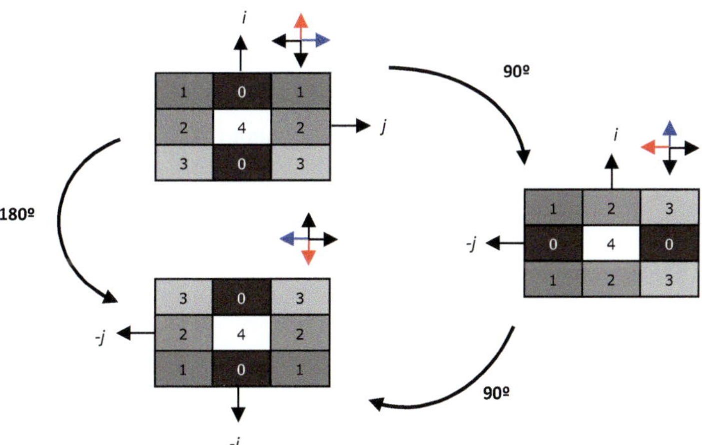

Fig. 5.11 Modeling of coordinate $H(-i, -j)$ inversion by 180° rotation of the filter coefficient matrix

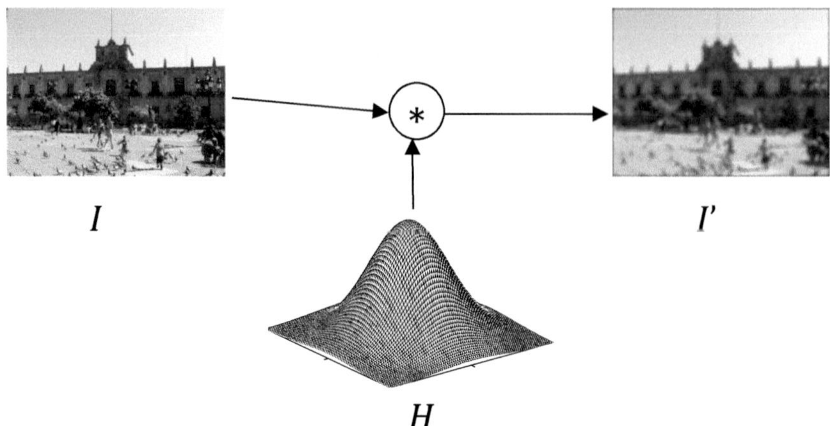

Fig. 5.12 Convolution using a Gaussian filter. The original image I is processed through convolution with the coefficient matrix $H(i,j)$ resulting in the transformed image I'

Closely related to convolution is the concept of correlation. In correlation, the filter mask $H(i,j)$ is applied to the image directly, without the $180°$ rotation of its coefficients.

Thus, while the two operations are nearly identical, the distinction lies in whether the filter matrix is flipped or not. In practical terms:

Correlation = filter applied as defined.
Convolution = same operation, but with the filter rotated by $180°$.

Regardless of this difference, the essential mechanism behind all linear filters is the convolution operation, denoted by the symbol (*). The specific effect of the filtering depends entirely on the chosen coefficients of the matrix $H(i,j)$. Figure 5.12 demonstrates the convolution process in the context of digital images.

Linear filters in Python

#Code 3.2

Difference filter

```python
#The libraries necessary for image processing are loaded
import cv2 as cv
import numpy as np
import matplotlib.pyplot as plt
from google.colab.patches import cv2_imshow

img = cv.imread('Figure0501.png')
blur = cv.blur(img,(9,9))
gauss = cv.GaussianBlur(img,(5,5),0)
gaussGray = cv.GaussianBlur(imgGray,(5,5),0)

# for image display
img = cv.imread('Figure0501.png')
cv2_imshow(img)
```

```python
# Difference Filter Mexican Hat
kernelDif = np.float32([[0,0,-1,0,0], [0,-1,-2,-1,0], [-1,-2,16,-2,-1], [0,-1,-2,-1,0], [0,0,-1,0,0]])
dif = cv.filter2D(img,-1,kernelDif)
cv2_imshow(dif)
```

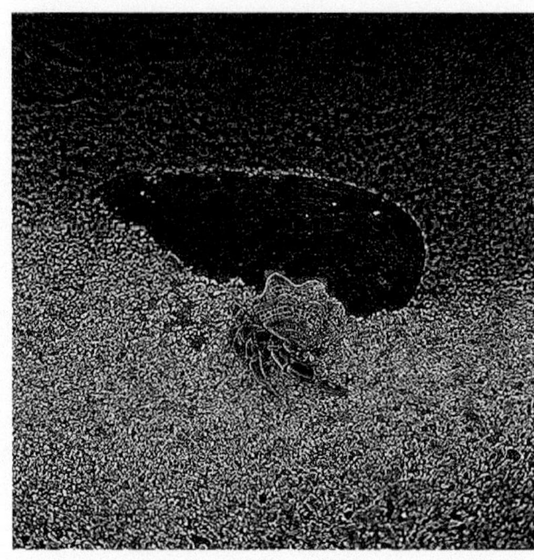

```
# Difference Filter Mexican Hat with filter2D on grayscale
image
difGray = cv.filter2D(imgGray,-1,kernelDif)
cv2_imshow(difGray)
```

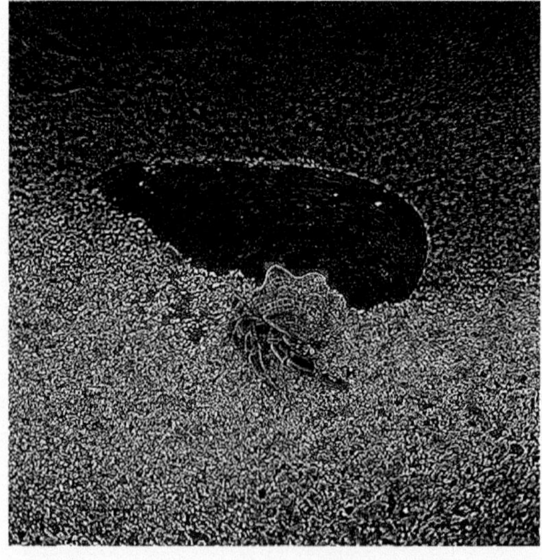

```
# Difference Filter Mexican Hat with filter2D on RGB image
gaussDif = cv.filter2D(gauss,-1,kernelDif)
cv2_imshow(gaussDif)
```

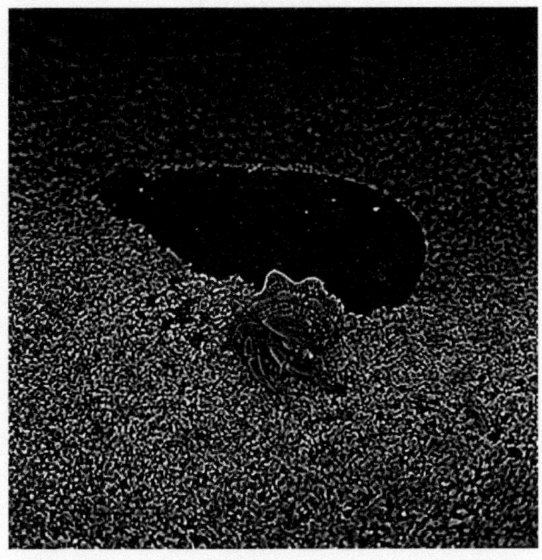

```
# Difference Filter Mexican Hat with filter2D on RGB image
gaussDifGray = cv.filter2D(gaussGray,-1,kernelDif)
cv2_imshow(gaussDifGray)
```

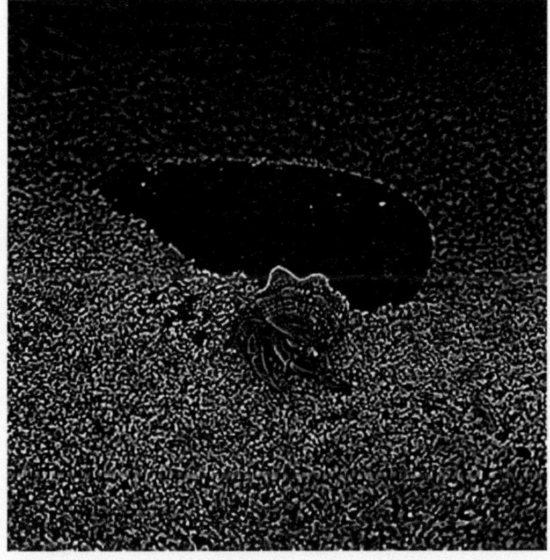

5.5.2 *Properties of Linear Convolution*

The significance of convolution lies in both its mathematical simplicity and its broad applicability in modeling real-world processes. Later chapters will also highlight its strong connection with Fourier analysis and frequency-domain methods. Below are the main properties that make convolution so fundamental.

Commutativity

Convolution is a commutative operation, which means the order of the operands does not matter:

$$I * H = H * I. \tag{5.14}$$

Whether the image I is convolved with the filter $H(i, j)$ or the filter is applied in reverse order, the outcome remains the same.

Linearity

Convolution is also linear, a property with several consequences for image processing:

Scaling: Multiplying an image by a constant a before convolution is equivalent to convolving first and then scaling:

$$(a \cdot I) * H = a \cdot (I * H). \tag{5.15}$$

Additivity: If two images are added pixel by pixel and then convolved, the result equals the sum of their individual convolutions:

$$(I_1 + I_2) * H = (I_1 * H) + (I_2 * H). \tag{5.16}$$

Offset limitation: Interestingly, adding a constant b to every pixel before convolution does not produce the same result as convolving first and then adding b:

$$(b + I) * H \neq b + I * H. \tag{5.17}$$

While these relationships are valuable theoretically, in practice, linearity can be affected by rounding errors or by constraints on image display values (commonly known as clamping).

Associativity

Convolution is associative, which means the grouping of multiple convolutions does not affect the final result:

$$A * (B * C) = (A * B) * C. \tag{5.18}$$

This property is important for filter design. For instance, a sequence of filters can be combined into a single equivalent filter, or conversely, one filter can be decomposed into several simpler ones.

5.5.3 Separability of Filters

A direct implication of associativity is the separability of specific filters. A large, complex filter can sometimes be expressed as the convolution of multiple smaller filters:

$$I * H = I * (H_1 * H_2 * H_3 \ldots H_n). \tag{5.19}$$

This decomposition is highly advantageous in practice. Even though the number of operations increases, each smaller filter is simpler and requires fewer computations, leading to a significant improvement in processing speed.

x–y Separation

An important optimization in filter design is the ability to separate a two-dimensional filter H into two simpler one-dimensional filters, H_x and H_y, which act horizontally and vertically on the image. Suppose we define two one-dimensional filters operating along each axis:

$$H_x = \begin{bmatrix} 1 & 1 & \underline{1} & 1 & 1 \end{bmatrix} \quad \text{or} \quad H_y = \begin{bmatrix} 1 \\ \underline{1} \\ 1 \end{bmatrix}. \tag{5.20}$$

If both filters are applied to an image I, the process can be expressed as:

$$I \leftarrow (I * H_x) * H_y = I * \underbrace{H_x * H_y}_{H_{xy}}. \tag{5.21}$$

Here, H_{xy} represents the two-dimensional filter obtained by convolving H_x and H_y:

$$H_{xy} = H_x * H_y = \begin{bmatrix} 1 & 1 & 1 & 1 & 1 \\ 1 & 1 & \underline{1} & 1 & 1 \\ 1 & 1 & 1 & 1 & 1 \end{bmatrix}. \tag{5.22}$$

In this case, the result is equivalent to the well-known Box filter, which averages neighboring pixels. However, the computational cost differs depending on how the filtering is performed. Applying the whole two-dimensional filter H_{xy} directly requires $3 \times 5 = 15$ operations per pixel. By separating it into two one-dimensional filters H_x and H_y, the same effect can be achieved with only $3 + 5 = 8$ operations per pixel. This example illustrates a broader principle: as the filter size increases, the number of required computations skyrockets. By separating filters into independent directions whenever possible, significant efficiency gains can be achieved without changing the result of the convolution.

Separation of the Gaussian Filter

As shown earlier, a two-dimensional filter can often be decomposed into two one-dimensional filters acting along the horizontal (x) and vertical (y) directions. In general, if a function is separable, it can be expressed as the product of two functions, each corresponding to a single dimension:

$$H_{xy}(i, j) = H_i(i) \cdot H_y(j). \tag{5.23}$$

A significant example is the Gaussian function. The two-dimensional Gaussian can be expressed as the product of two independent one-dimensional Gaussians:

$$G_\sigma(x, y) = e^{-\frac{x^2+y^2}{2\sigma^2}} = e^{-\frac{x^2}{2\sigma^2}} \cdot e^{-\frac{y^2}{2\sigma^2}} = g_\sigma(x) \cdot g_\sigma(y). \tag{5.24}$$

This property means that a two-dimensional Gaussian filter, $H^{G,\sigma}$ can be implemented as two one-dimensional filters $H_x^{G,\sigma}$ and $H_y^{G,\sigma}$ applied sequentially:

$$I \leftarrow I * H^{G,\sigma} = I * H_x^{G,\sigma} * H_y^{G,\sigma}. \tag{5.25}$$

This separability offers significant computational benefits. A direct two-dimensional Gaussian filter requires a large mask, especially for higher values of the standard deviation σ. For example, with $\sigma = 10$, a 51×51 filter would be needed. However, by separating it into two one-dimensional operations, the same smoothing effect can be achieved with dramatically fewer computations—in this case, up to 50 times faster than using the whole two-dimensional filter directly.

The Gaussian filter's gradual decay also means that minimal values of $\sigma < 2.5$ are not practical, as rounding errors in the coefficients would compromise the accuracy of the result.

5.5.4 Filter Impulse Response

In convolution, there exists a neutral element, known as the impulse function or the Dirac delta function $\delta()$. Its defining property is:

$$I * \delta = I. \tag{5.26}$$

This means that convolving an image with the impulse function leaves the image completely unchanged.

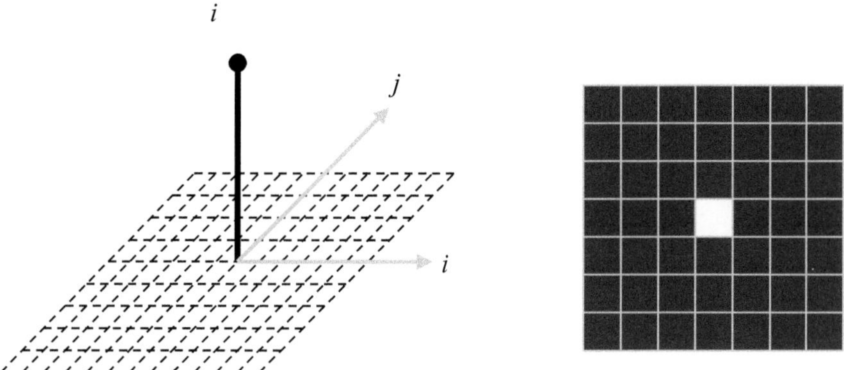

Fig. 5.13 Two-dimensional representation of the Dirac delta function

In the discrete two-dimensional case, the delta function is defined as:

$$\delta(i,j) = \begin{cases} 1 \text{ for } i = j = 0 \\ 0 \text{ else} \end{cases}.$$ (5.27)

If interpreted as an image, the Dirac delta function appears as a single bright pixel at the origin surrounded by an infinite field of black pixels. Figure 5.13 illustrates this visualization. When the delta function is used as a filter, applying convolution to an image yields the original image itself, as shown in Fig. 5.14. Conversely, when the delta function is treated as the input image and convolved with a filter H, the result is the filter's coefficient matrix:

$$H * \delta = H.$$ (5.28)

This second property is especially significant. It allows us to determine the characteristics of an unknown filter: by applying the impulse function to it, the filter's structure is revealed. The resulting output is known as the impulse response of the filter.

The impulse response is a central concept in both image processing and engineering system analysis, as it provides a direct way to describe, measure, and understand the behavior of linear filters (Fig. 5.15).

5.6 Spatial Nonlinear Filters

Linear filters are widely used for tasks such as smoothing and noise reduction. However, they come with a significant limitation: important image structures—such as points, edges, and fine lines—are also blurred or weakened in the process

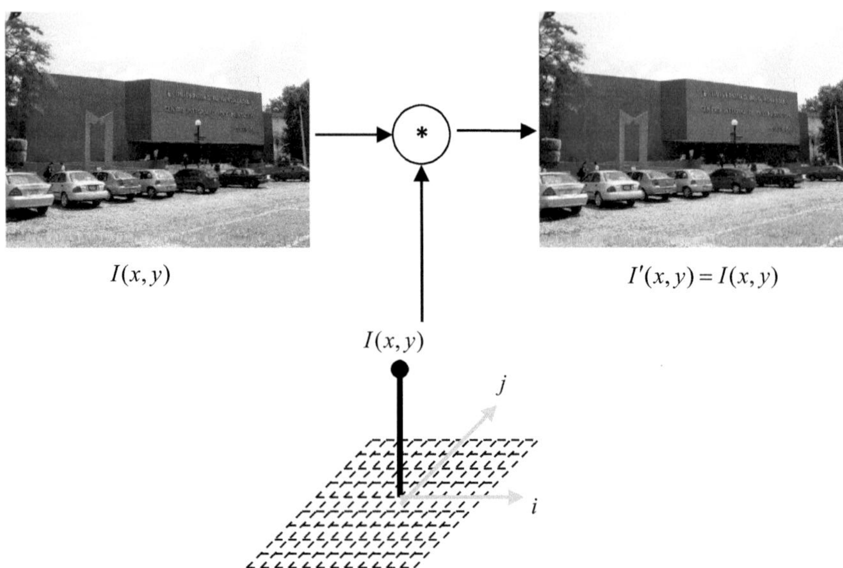

Fig. 5.14   Output obtained when the Dirac delta function δ is convolved with an image I

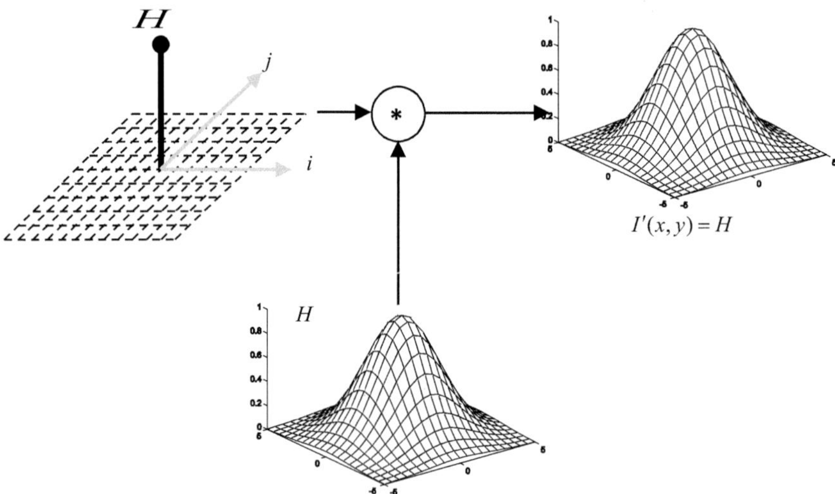

Fig. 5.15   Result of convolving the impulse function δ with a filter H

(see Fig. 5.16). No arrangement of linear filter coefficients can altogether avoid this drawback.

When the goal is to reduce noise without degrading structural details, linear filters quickly reach their limits. To address this issue, a different approach is required: the use of nonlinear filters.

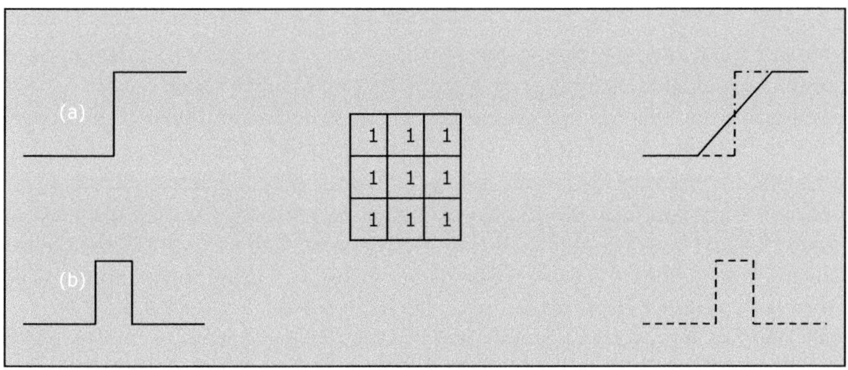

Fig. 5.16 Demonstration of how linear smoothing influences image details, highlighting the blurring of step boundaries (**a**) and the weakening of line structures (**b**)

Nonlinear filters operate according to principles that do not adhere to linear mathematical properties. By breaking away from linearity, they can smooth homogeneous regions of an image while preserving critical details, offering a more effective solution for many image enhancement tasks.

5.6.1 Maximum and Minimum Filters

Nonlinear filters, similar to linear ones, compute the value of a pixel at position (x, y), l using information from a neighborhood region $R(x, y)$ around it. Among the simplest nonlinear operations are the minimum and maximum filters, defined as:

$$I'(x, y) = \min\{I(x + i, y + j)|(i, j) \in R\} = \min(R(x, y))$$
$$I'(x, y) = \max\{I(x + i, y + j)|(i, j) \in R\} = \max(R(x, y)). \qquad (5.29)$$

Here, $R(x, y)$ is the local window centered at (x, y) which is typically a small rectangle such as 3×3. When the minimum filter is applied to an image, step edges tend to shift toward the right by an amount proportional to the window's width, while thin bright lines—narrower than the filter's span—can disappear completely. Conversely, the maximum filter produces the opposite effect by shifting edges in the other direction and reinforcing bright structures. These filters are handy for handling salt-and-pepper noise, where random white (salt) and black (pepper) pixels are added to an image. The minimum filter is effective at removing salt noise: when centered on a bright outlier, it replaces the pixel with the lowest intensity found in its neighborhood. However, this also enlarges dark regions, since black pixels are preserved and can spread according to the filter size. The maximum filter acts in reverse—suppressing pepper noise (dark pixels) while strengthening white regions in the image.

Figures 5.17 and 5.18 illustrate these effects. Figure 5.17 illustrates how the minimum filter affects image edges and fine lines, while Fig. 5.18 demonstrates both filters applied to an image corrupted by salt-and-pepper noise.

Illustrates the effect of applying minimum and maximum filters to an intensity image deliberately degraded with salt-and-pepper noise. This type of noise introduces random bright pixels ("salt") and dark pixels ("pepper") across the scene. The minimum filter suppresses the bright disturbances by replacing the value of each pixel with the lowest intensity found within its neighborhood region $R(x, y)$. Consequently, isolated white pixels are substituted by darker values from their surroundings, while black pixels tend to expand according to the filter size A.

In contrast, the maximum filter performs the inverse operation. It removes or reduces dark noise components, such as pepper pixels or other low-intensity values,

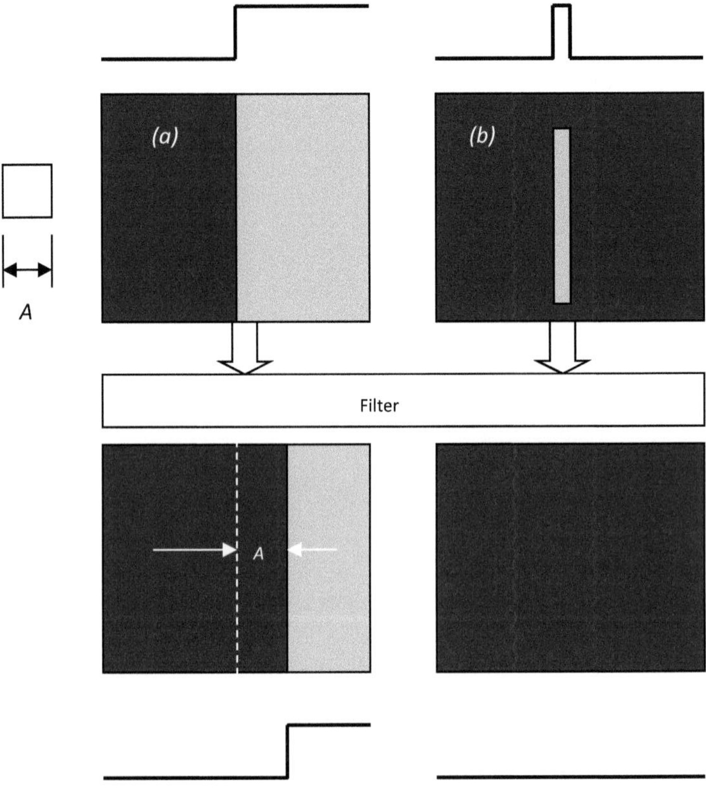

Fig. 5.17 Impact of the minimum filter on various local image structures. The upper row shows the original patterns, while the lower row displays the results after applying the filter. Parameter A indicates the size of the filter, which also specifies the extent of the neighborhood region R. In case **a**, the step edge is displaced to the right by a distance equal to A. In case **b**, if the line is narrower than the filter with A, it vanishes completely under the action of the filter

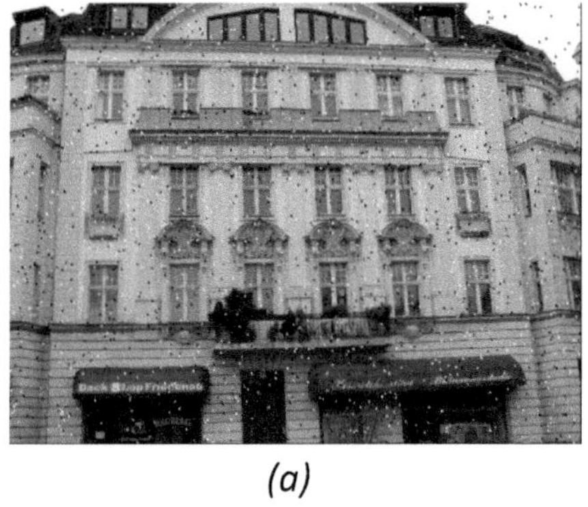

(a)

(b)   (c)

Fig. 5.18 Use of minimum and maximum filters. **a** Original image affected by artificially added salt-and-pepper noise, **b** result produced with a 3×3 minimum filter, and **c** result obtained with a 3×3 maximum filter

while simultaneously strengthening white structures, thereby enhancing the lighter details of the image.

5.6.2 The Median Filter

No filter is capable of simultaneously eliminating all types of noise from an image while preserving every detail that might be considered important. This limitation arises because a filter cannot distinguish which structures the observer values and which can be discarded. The median filter provides a practical balance between noise removal and structure preservation.

Unlike linear smoothing filters, which often blur and weaken image edges, the median filter effectively reduces noise and minor artifacts without severely degrading boundary information. It belongs to the category of statistical filters, characterized by nonlinear operations that rely on statistical measures.

In statistics, the median is defined as the value that divides a dataset into two halves: 50% of the values are less than or equal to it, and the remaining 50% are greater than or equal to it. If we consider a sample of value $x_1, x_2, \ldots, x_n$, arranged in ascending order, the median is formally given by:

$$M_e = x_{\frac{n+1}{2}}. \tag{5.30}$$

When n is odd, it means the central observation of the ordered data. If n is even, the median is the arithmetic mean of the two central values:

$$M_e = \frac{x_{\frac{n}{2}} + x_{\frac{n+1}{2}}}{2} \tag{5.31}$$

Based on this definition, the median filter replaces each pixel in an image with the median intensity value within its neighborhood region $R(x, y)$, as expressed by:

$$I'(x, y) = M_e(R(x, y)). \tag{5.32}$$

The procedure for computing the median in $R(x, y)$ involves two simple steps: (1) collect the intensity values within the defined region and arrange them into a vector, and (2) reorder these values in ascending order. Repeated values remain duplicated in this ordered list (Fig. 5.19).

Unlike other filters that compute a new replacement value, the median filter selects one of the existing intensity values from within the neighborhood, chosen after the local data have been ordered. Figure 5.20 illustrates its action on two-dimensional shapes: when the structures are smaller than half the filter's window size, as in

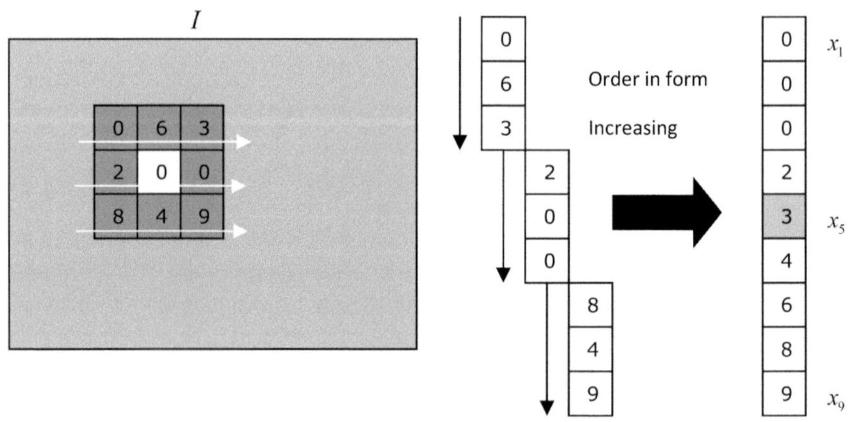

Fig. 5.19 Process of calculating the median for a filter region of size 3 × 3

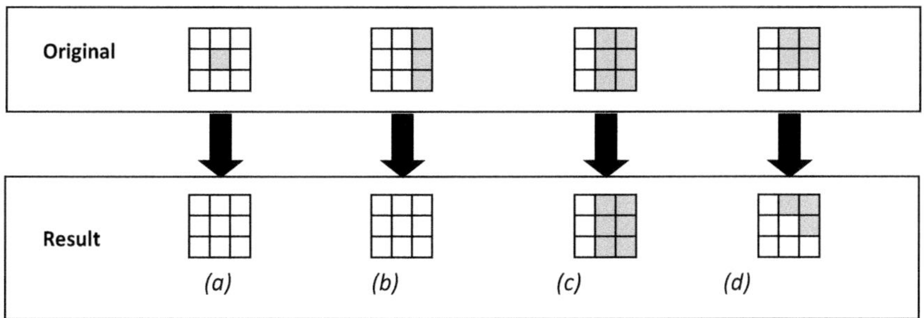

Fig. 5.20 Effect of the median filter on different structures. In cases **a** and **b**, structures smaller than half the filter size are removed, while in **c** and **d**, structures equal to or larger than half the filter size remain preserved mainly

Fig. 5.20a and b, they are eliminated; when their size is equal to or greater than half the window, as shown in Fig. 5.20c and d, they remain almost unchanged.

Figure 5.21 further highlights the advantages of this method, showing that the median filter effectively removes salt-and-pepper noise while maintaining edge definition, in contrast to a smoothing filter, which only reduces the artifacts but fails to eliminate them.

5.6.3 The Median Filter with Multiplicity Window

The conventional median filter determines its result by ranking the values within the neighborhood region $R(x, y)$. Extreme values—whether very high or very low—do not significantly affect the outcome, making the method particularly effective for removing salt-and-pepper noise. In its standard form, however, all pixels in the neighborhood contribute equally to the calculation. Ideally, though, the central pixel (the one being replaced) should have a greater influence than its neighbors.

The median filter with a multiplicity window is a variant designed to address this. Instead of treating each pixel equally, it assigns different weights according to the coefficients of a filter matrix H. Similar to linear filters, these coefficients scale the contribution of each pixel. The difference lies in the operation: instead of multiplying the pixel intensity by the coefficient, the coefficient indicates how many times that pixel's value is repeated in the dataset from which the median is determined.

Thus, the sum of all coefficients in the matrix defines the total number of data points T considered in the median calculation. For a 3×3 window with a coefficient matrix:

$$H = \begin{bmatrix} h_1 & h_2 & h_3 \\ h_4 & h_5 & h_6 \\ h_7 & h_8 & h_9 \end{bmatrix}. \tag{5.33}$$

(a)

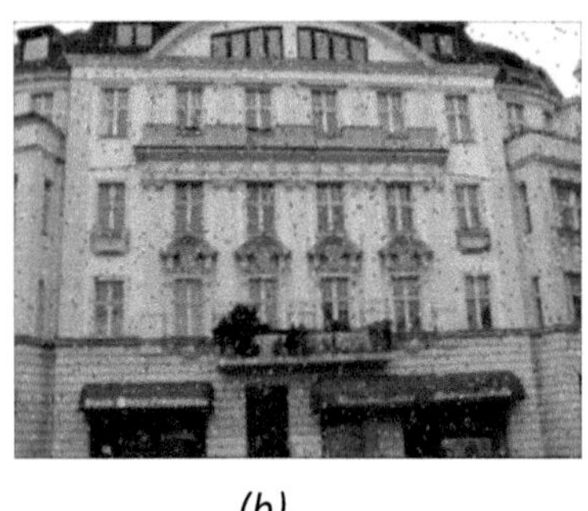

(b)

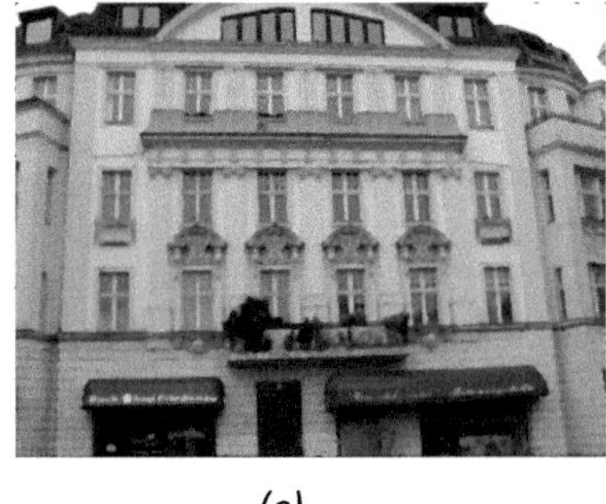

(c)

Fig. 5.21 Comparison of the median filter with a smoothing filter. **a** Original image corrupted with salt-and-pepper noise. **b** Result after applying a box smoothing filter, where noise artifacts are only attenuated but not eliminated. **c** Result after applying the median filter, showing effective suppression of the noise while maintaining the integrity of the image

The number of values used is:

$$T = h_1 + h_2 + h_3 + h_4 + h_5 + h_6 + h_7 + h_8 + h_9 = \sum_{i=1}^{9} h_i. \qquad (5.34)$$

For example, if the filter is defined as:

$$H = \begin{bmatrix} 1 & 3 & 2 \\ 2 & \underline{4} & 1 \\ 1 & 2 & 1 \end{bmatrix} \qquad (5.35)$$

then the number of elements considered is:

$$T = \sum_{i=1}^{9} h_i = 17. \qquad (5.36)$$

The rule for determining the median remains the same: if T is odd, the median corresponds to the central element in the ordered dataset; if T is even, it is the average of the two central values, as described in Eq. (5.29)

Figure 5.22 Operation of the median filter with a multiplicity window.

To ensure proper functioning, all coefficients of H must be non-negative. A zero coefficient means that the corresponding pixel is excluded from the median calculation. A typical example is the median cross filter, defined as:

$$H = \begin{bmatrix} 0 & 1 & 0 \\ 1 & \underline{1} & 1 \\ 0 & 1 & 0 \end{bmatrix}.$$

(5.37)

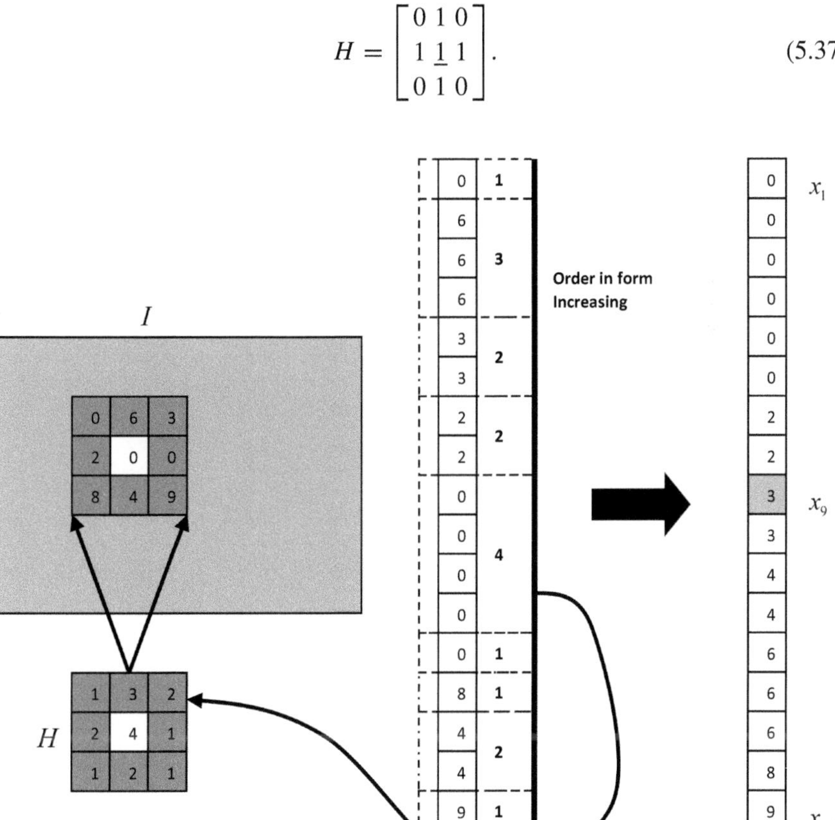

Fig. 5.22 Calculation developed by the median filter with a window of multiplicity of 3×3

Nonlinear Filters in Python

#Code 3.3

 # Difference filter

```
#The libraries necessary for image processing are loaded
import cv2 as cv
import numpy as np
import matplotlib.pyplot as plt
from google.colab.patches import cv2_imshow # for image display

img = cv.imread('Figure0511.png')
cv2_imshow(img)
```

```
# for grayscale image display Add noise, salt and pepper
imgGray = cv.cvtColor(img,cv.COLOR_BGR2GRAY)
cv2_imshow(imgGray)
```

```
# Add noise, salt and pepper
#Add salt and pepper noise to image with prob as the
probability of the noise
def sp_noise(image, prob):
    output = image.copy()
    if len(image.shape) == 2: #Grayscale image
        black = 0
        white = 255
    else: # RGB
      black = np.array([0, 0, 0], dtype='uint8')
      white = np.array([255, 255, 255], dtype='uint8')
    probs = np.random.random(output.shape[:2])#Generates N by M
random probabilities, where N and M are computed with
shape[:2], which takes the first two dimensions of the image
    output[probs < prob] = black
    output[probs > 1 - prob] = white
    return output
imgNoise = sp_noise(img,0.01)
cv2_imshow(imgNoise)
```

```python
# Minimum filter
def minFilter(imgNoise):
  im = imgNoise.copy()
  vector = []
  rows, cols = im.shape[:2]
  for i in range(rows-2):
    i += 1
    for j in range(cols-2):
      j += 1
      vector.append(imgNoise[i-1, j-1])
      vector.append(imgNoise[i - 1, j])
      vector.append(imgNoise[i-1, j + 1])
      vector.append(imgNoise[i, j - 1])
      vector.append(imgNoise[i, j])
      vector.append(imgNoise[i, j + 1])
      vector.append(imgNoise[i + 1, j - 1])
      vector.append(imgNoise[i + 1, j])
      vector.append(imgNoise[i + 1, j + 1])
      min = np.min(vector,axis=0)
      im[i, j,:] = min
      vector = []
  return im

imgMin = minFilter(imgNoise)
cv2_imshow(imgMin)
```

```python
# Maximum filter
def maxFilter(imgNoise):
  im = imgNoise.copy()
  vector = []
  [rows, cols] = im.shape[:2]
  for i in range(rows-2):
    i += 1
    for j in range(cols-2):
      j += 1
      vector.append(imgNoise[i-1, j-1])
      vector.append(imgNoise[i - 1, j])
      vector.append(imgNoise[i-1, j + 1])
      vector.append(imgNoise[i, j - 1])
      vector.append(imgNoise[i, j])
      vector.append(imgNoise[i, j + 1])
      vector.append(imgNoise[i + 1, j - 1])
      vector.append(imgNoise[i + 1, j])
      vector.append(imgNoise[i + 1, j + 1])
      max = np.max(vector,axis=0)
      im[i, j] = max
      vector = []
  return im

imgMax = maxFilter(imgNoise)
cv2_imshow(imgMax)
```

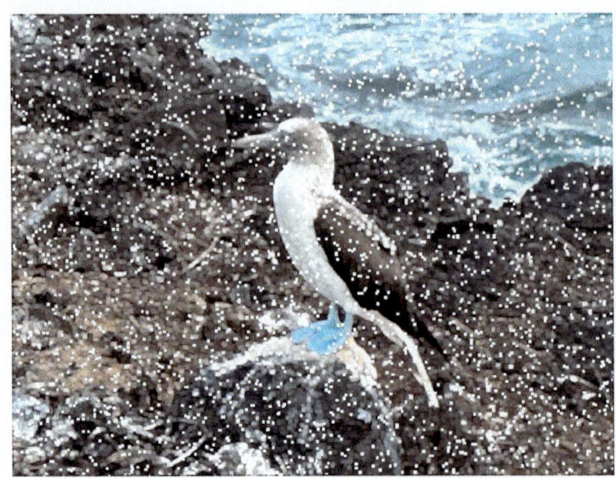

```python
# Median filter
def medianFilter(imgNoise):
  im = imgNoise.copy()
  vector = []
  [rows, cols] = im.shape[:2]
  for i in range(rows-2):
    i += 1
    for j in range(cols-2):
      j += 1
      vector.append(imgNoise[i-1, j-1])
      vector.append(imgNoise[i - 1, j])
      vector.append(imgNoise[i-1, j + 1])
      vector.append(imgNoise[i, j - 1])
      vector.append(imgNoise[i, j])
      vector.append(imgNoise[i, j + 1])
      vector.append(imgNoise[i + 1, j - 1])
      vector.append(imgNoise[i + 1, j])
      vector.append(imgNoise[i + 1, j + 1])
      median = np.median(vector,axis=0)
      im[i, j] = median
      vector = []
  return im

imgMedian = medianFilter(imgNoise)
cv2_imshow(imgMedian)
```

```
# Median filter with medianBlur
imgMedianBlur = cv.medianBlur(imgNoise,3)
cv2_imshow(imgMedianBlur)
```

```
plt.figure(figsize=(15, 5))
plt.subplot(1,3,1)
plt.imshow(cv.cvtColor(imgNoise, cv.COLOR_BGR2RGB))
plt.title('Image with noise')
plt.subplot(1,3,2)
plt.imshow(cv.cvtColor(imgMedian, cv.COLOR_BGR2RGB))
plt.title('imgMedian')
plt.subplot(1,3,3)
```

```
plt.imshow(cv.cvtColor(imgMedianBlur, cv.COLOR_BGR2RGB))
plt.title('imgMedianBlur')
plt.show()
```

5.6.4 Other Nonlinear Filters

The median filter and its multiplicity window variant are just two representative examples of nonlinear filtering techniques, chosen for their simplicity and frequent use. By definition, the category of *nonlinear filters* includes all filters that lack the property of linearity. Within this broad class, many specialized approaches exist, such as corner detection filters and morphological filters, which will be addressed in later chapters.

A key distinction between linear and nonlinear filters lies in their theoretical foundations. Linear filters are firmly grounded in convolution theory, which provides a well-established mathematical framework and a wide range of inherent properties. Nonlinear filters, on the other hand, do not share this everyday basis, and their underlying theory is comparatively less unified and less developed.

References

1. Gonzalez, R. C., & Woods, R. E. (2018). *Digital image processing* (4th ed.). Pearson.
2. Jähne, B. (2005). *Digital image processing: Concepts, algorithms, and scientific applications* (6th ed.). Springer.
3. Pratt, W. K. (2007). *Digital image processing: PIKS inside* (4th ed.). Wiley-Interscience.
4. Russ, J. C. (2011). *The image processing handbook* (6th ed.). CRC Press.
5. Tomasi, C., & Manduchi, R. (1998). Bilateral filtering for gray and color images. In: *Proceedings of the Sixth IEEE International Conference on Computer Vision (ICCV'98)* (pp. 839–846).

Chapter 6
Edges and Contours

Edges and contours constitute essential low-level features in image analysis, typically extracted through the detection of local gradients or abrupt changes in intensity or color. These features often delineate object boundaries and structural transitions, providing critical information for higher-level image interpretation tasks [1].

The perceived clarity or definition of an image is strongly correlated with the presence and sharpness of these discontinuities. From a computational perspective, sharp edges enhance the detectability of structures within an image, improving both segmentation accuracy and object recognition performance.

Moreover, the human visual system is susceptible to edge information. Even minimal edge representations, such as line drawings, can convey sufficient information for object identification. This perceptual sensitivity underscores the importance of accurate edge and contour detection in fields such as computer vision, image processing, and pattern recognition.

This chapter provides an in-depth overview of the primary edge localization techniques, examining their theoretical foundations, implementation strategies, and typical use cases in various imaging applications.

6.1 How Are Contours Produced?

Edges perform a central role in human vision and are likely fundamental in other biological visual systems as well. Their perceptual salience allows humans to recognize and reconstruct objects even when presented with minimal visual information. Often, just a few edge lines are sufficient to form a coherent mental representation of complex shapes (see Fig. 6.1). This ability underscores the significance of edge information as a concise and efficient descriptor of visual structures.

From a computational standpoint, understanding how edges originate and how they can be detected algorithmically is a key challenge in image analysis. Edges are

© The Author(s), under exclusive license to Springer Nature Switzerland AG 2026

E. Cuevas et al., *Image Processing with Python*, Signals and Communication Technology, https://doi.org/10.1007/978-3-032-13285-7_6

(a)   (b)

Fig. 6.1 **a** Grayscale image, and **b** detected edges

typically defined as locations in an image where there is a significant variation in intensity over a localized region, particularly along a specific direction. These abrupt intensity transitions correspond to high spatial frequency components, which can be effectively detected using differential operators [2].

In practice, the magnitude of the intensity change, often referred to as the gradient magnitude, is computed using first-order derivatives. These derivatives quantify the rate of intensity change across the image and serve as the foundation for many edge detection algorithms. The larger the gradient at a given pixel, the more likely it is that the pixel lies on an edge. Thus, differential-based approaches are among the most widely used and theoretically grounded methods for edge detection in digital images.

6.2 Methodologies Based on the Gradient for Edge Detection

To facilitate understanding, consider a simplified one-dimensional case. Suppose an image contains a bright (white) region centered within a darker background, as illustrated in Fig. 6.2a. Suppose we extract a horizontal scan line across this image. In that case, the corresponding grayscale intensity profile (represented as a function (u), where u denotes the spatial coordinate along the scan line) might resemble the profile shown in Fig. 6.2b.

This one-dimensional signal serves as a valuable model for analyzing edge behavior in a controlled setting. To identify the presence and location of edges, we compute the first derivative $f'(u)$ of the signal, see Eq. 6.1. These derivative measures the rate of change in intensity for the spatial position. At points where the intensity transitions sharply, such as from bright to dark or dark to bright, the derivative exhibits high magnitude values, typically manifesting as peaks or valleys

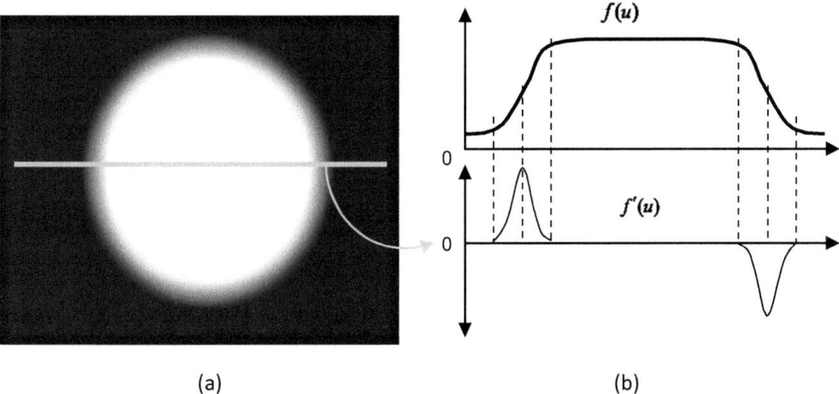

Fig. 6.2 Example in one dimension of the first derivative considering the x axis information. **a** Grayscale image and **b** the one-dimensional function $f(u)$ and its derivative $f'(u)$

in the derivative signal. These peaks in the first derivative are strong indicators of edge locations. As such, this method forms the basis for many gradient-based edge detection techniques.

$$f'(u) = \frac{df}{du}(u). \tag{6.1}$$

This results in a negative change, where the intensity diminishes, and a positive variation, where the intensity rises. Nevertheless, it is necessary to use a special technique to calculate derivatives, as they are not defined for discrete functions.

The **rate of change** of a continuous function at a specific point x is described by its **derivative** at that point. This value represents the **slope of the tangent line** to the function's graph at the x axis, providing a measure of how the function behaves locally around that position. But, in digital image processing, image data is inherently discrete, consisting of sampled intensity values at specific intervals. In this context, a direct computation of the continuous derivative is not possible. Instead, we resort to numerical approximations of the derivative using finite differences.

To estimate the rate of change (i.e., the slope of the tangent) at a point u, we compute the difference between neighboring intensity values around u, divided by the sampling interval, as described in Eq. 6.2. This leads to the general difference approximation of the derivative.

$$\frac{df}{du}(u) \approx \frac{f(u+1) - f(u-1)}{2} = 0.5 \cdot (f(u+1) - f(u-1)). \tag{6.2}$$

A similar procedure can be carried out in a vertical direction along the image column.

6.2.1 *Calculation of Gradient for Edge Detection*

A partial derivative refers to the rate of change of a function for one of its variables, while keeping the other variables constant. In the context of image processing, where an image could be considered as a bivariate intensity function $I(x, y)$, the partial derivatives describe how the intensity changes along each spatial axis. The formal definition of these derivatives is given in Eq. 6.3.

The partial derivative of $I(x, y)$, considering the x axis, measures the rate of change in the horizontal direction (i.e., along image rows). Similarly, the partial derivative, considering the y axis, captures the vertical rate of change (i.e., along image columns).

These derivatives are critical for detecting edges in two-dimensional images, as edges are typically characterized by abrupt intensity changes along one or more directions.

$$\frac{\partial I}{\partial x}(x, y) \quad \text{and} \quad \frac{\partial I}{\partial y}(x, y). \tag{6.3}$$

Figure 6.3 illustrates how the derivative can be graphically interpreted in a one-dimensional discrete function. The slope of the line at points $(u - 1)$ and $(u + 1)$ serves as an indirect calculation of the slope to the tangent at $f(u)$. The calculation is sufficiently approximate in most cases that might arise.

By computing the partial derivatives along both axes, we can construct the gradient vector expressed in Eq. 6.4. This equation shows the partial derivative of the image function $I(x, y)$ considering the variable x or y.

$$\nabla I(x, y) = \begin{bmatrix} \frac{\partial I}{\partial x}(x, y) \\ \frac{\partial I}{\partial y}(x, y) \end{bmatrix}. \tag{6.4}$$

The magnitude, defined in Eq. 6.5, is commonly used as a measure of edge strength.

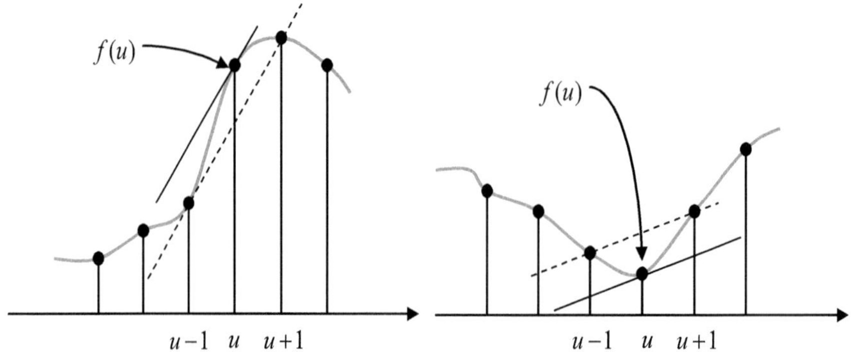

Fig. 6.3 Graphical interpretation of the derivative in a one-dimensional discrete function

$$|\nabla I| = \sqrt{\left(\frac{\partial I}{\partial x}\right)^2 + \left(\frac{\partial I}{\partial y}\right)^2}.$$ (6.5)

One of the key properties of the gradient magnitude in image processing is its invariance to image rotation. That is, the value of the gradient $|\nabla I|$ remains unchanged regardless of the orientation of the underlying structures in the image. This rotation invariance makes the gradient magnitude a robust and reliable measure for identifying significant intensity transitions, regardless of the direction in which they occur.

This characteristic is particularly valuable for localizing edge points, as edges can be oriented arbitrarily within the image plane. By relying on the gradient magnitude, edge detection algorithms can effectively identify edges without needing to account for their orientation explicitly. Due to this invariance and its strong correlation with edge strength, the gradient magnitude is widely used in edge detection algorithms as a primary measure for identifying edge locations.

6.2.2 Derivative Kernel

The elements of the gradient vector, as defined in Eq. 6.4, correspond to the partial derivatives of the image intensity function along the row x and column y directions. These derivatives quantify how the image intensity changes in the x and y axes, respectively, and are the foundation for computing directional gradients essential to edge detection.

As described in Eq. 6.2 and illustrated in Fig. 6.3, the horizontal derivative at a pixel location $I(x, y)$ can be approximated using a central difference scheme. This method estimates the rate of intensity change along the horizontal axis by subtracting the intensity of the left neighbor from that of the right neighbor and then dividing by two. This approximation can be compactly expressed as a convolution with the filter kernel of Eq. 6.6.

$$H_x^D = \left[-0.5 \ \underline{0} \ 0.5\right] = 0.5 \cdot \left[-1 \ \underline{0} \ 1\right].$$ (6.6)

In this kernel, the underscore symbol ("_") denotes the reference pixel (x), the center of the kernel. The coefficient -0.5 multiplies the left neighbor $(x - 1)$, and 0.5 multiplies the right neighbor $(x + 1)$. The central pixel is effectively ignored or assigned a weight of zero. This operation approximates the derivative at that pixel.

Analogously, the vertical gradient can be estimated by applying a similar kernel transposed for the vertical direction as described in Eq. 6.7.

$$H_y^D = \begin{bmatrix} -0.5 \\ \underline{0} \\ 0.5 \end{bmatrix} = 0.5 \cdot \begin{bmatrix} -1 \\ \underline{0} \\ 1 \end{bmatrix}.$$ (6.7)

This filter estimates the vertical derivative by taking the intensity difference between the pixel below and the one above the current location.

Figure 6.4 illustrates the effect of applying these directional gradient filters to an input image. In Fig. 6.4b, the horizontal gradient filter H_x^D yields a strong response at vertical edge regions where the intensity varies significantly along the horizontal axis. This is because vertical transitions (such as object boundaries running from top to bottom) manifest as strong horizontal gradients. Similarly, in Fig. 6.4c, the vertical gradient filter H_y^D produces prominent responses at horizontal edges, those with strong intensity changes along the vertical axis. This directional selectivity makes these filters particularly useful for detecting edge orientation and structure.

In regions of the image where intensity changes are minimal or absent in the corresponding direction, the filter responses are close to zero. In Fig. 6.4b and c, these regions are visually represented using gray pixels, indicating null or negligible gradient magnitude in the respective directions.

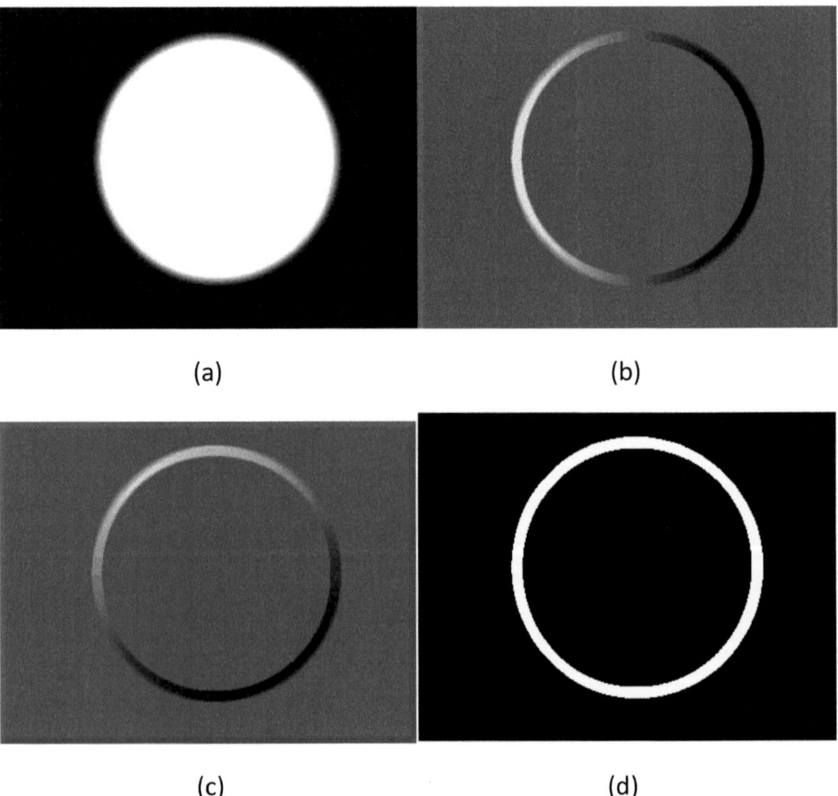

(a)   (b)

(c)   (d)

Fig. 6.4 Illustration of the image derivative. **a** Original image, **b** the derivative over the axis $\partial I/\partial x$, **c** over the axis $\partial I/\partial y$, and **d** the gradient $|\nabla I|$. In images (**b**) and (**c**), the bright zones correspond to positive values, and the dark zones to negative values. The gray zones refer to zero values

6.3 Filters for Edge Detection

Edge detection filters are designed to estimate the local gradient of an image, and the way this gradient is computed fundamentally characterizes each edge detection operator. While the overarching goal is the same (to detect intensity discontinuities indicative of object boundaries), the specific implementation details vary among operators.

The key differences among these methods lie in how the gradient components in the horizontal and vertical directions are approximated, as well as in how these directional gradients are combined into a single measure of edge strength. Some operators use simple differences, while others apply more complex convolution masks that incorporate smoothing or noise suppression.

In many applications, it is desirable to extract not only the magnitude of the gradient (which correlates with the presence and strength of an edge) but also the orientation (direction) of the edge. Fortunately, both pieces of information are inherently contained within the computed gradient vector at each pixel. The gradient magnitude provides a scalar value that indicates the strength of the edges. In contrast, the gradient direction, usually expressed as an angle θ, indicates the orientation perpendicular to the edge [3].

Due to the practical importance of gradient-based edge information, various edge detection operators have been developed, each with its advantages in terms of accuracy, noise sensitivity, and computational complexity. Some of the most widely known and historically significant operators, such as the Roberts, Prewitt, Sobel, and Canny detectors, are discussed in the following sections. These operators have become standard tools in digital image processing and computer vision due to their effectiveness in diverse applications ranging from medical imaging to autonomous navigation.

6.3.1 Operators Prewitt and Sobel

The **Prewitt** and **Sobel** operators are two of the most widely used gradient-based edge detection methods in digital image processing. Their popularity stems from their simplicity, efficiency, and relatively good performance in detecting edges under various conditions. While both methods follow the same principle (estimating the first-order derivatives of the image intensity function using convolution kernels), they differ slightly in how they assign weights to neighboring pixels, which affects their behavior in the presence of noise.

The Filter

Both operators use 3×3 convolution masks to approximate the image gradient in horizontal and vertical directions. This kernel size allows them not only to estimate local intensity differences but also to incorporate a smoothing effect, which is essential for reducing the influence of high-frequency noise.

In contrast to the simpler difference operators described in Eqs. 6.6 and 6.7, the Prewitt and Sobel filters are more robust to noise because they effectively combine derivative computation with local averaging.

The Prewitt operator uses the kernels described in Eq. 6.8 to compute the gradient in horizontal and vertical directions.

$$H_x^P = \begin{bmatrix} -1 & 0 & 1 \\ -1 & 0 & 1 \\ -1 & 0 & 1 \end{bmatrix} \quad \text{and} \quad H_y^P = \begin{bmatrix} -1 & -1 & -1 \\ 0 & 0 & 0 \\ 1 & 1 & 1 \end{bmatrix}. \tag{6.8}$$

Each filter is convolved with a 3×3 region centered at the pixel of interest, and the resulting response indicates the intensity change in the respective direction. These masks are designed to detect edges by enhancing regions of high intensity variation while applying a small amount of smoothing simultaneously.

The Prewitt kernels can also be interpreted as the product of two 1D filters (see Eq. 6.9), which highlights the roles of derivative and smoothing components.

$$H_x^P = \begin{bmatrix} 1 \\ 1 \\ 1 \end{bmatrix} * \begin{bmatrix} -1 & 0 & -1 \end{bmatrix} \quad \text{or} \quad H_y^P = \begin{bmatrix} 1 & 1 & 1 \end{bmatrix} * \begin{bmatrix} -1 \\ 0 \\ -1 \end{bmatrix}. \tag{6.9}$$

This separable form reveals the two essential operations that the Prewitt operator performs. Derivative estimation: The vectors $[-1\ 0\ -1]$ and $[-1\ 0\ -1]^T$ approximate the first-order derivative along the x and y axes, respectively. These are consistent with the different methods introduced earlier in Sect. 6.2.2.

The accompanying vectors $[1\ 1\ 1]^T$ and $[1\ 1\ 1]$ act as averaging filters in the direction orthogonal to the derivative. This helps to mitigate the effect of noise by reducing random variations before the derivative is applied.

Thus, the Prewitt filter not only detects edges but also incorporates a built-in smoothing operation, improving robustness to noise. This dual effect (gradient estimation with local averaging) makes it a practical tool for real-world edge detection tasks, particularly in moderately noisy images.

**Note**

The Sobel and Prewitt operators are almost identical filters. However, the key difference lies in how the Sobel operator assigns greater weight to the center row or column of the filter, depending on the direction of the gradient being computed. The kernel that represents the Sobel filter is stated in Eq. 6.10.

$$H_x^S = \begin{bmatrix} -1 & 0 & 1 \\ -2 & 0 & 2 \\ -1 & 0 & 1 \end{bmatrix} \quad \text{and} \quad H_y^S = \begin{bmatrix} -1 & -2 & -1 \\ 0 & 0 & 0 \\ 1 & 2 & 1 \end{bmatrix}. \tag{6.10}$$

These filters estimate the local gradient of the pixels in the image in both x and y dimensions, preserving the relation shown in Eq. 6.11.

$$\nabla I(x, y) \approx \frac{1}{6}\begin{bmatrix} H_x^P \cdot I \\ H_y^P \cdot I \end{bmatrix} \quad \text{and} \quad \nabla I(x, y) \approx \frac{1}{8}\begin{bmatrix} H_x^S \cdot I \\ H_y^S \cdot I \end{bmatrix}. \tag{6.11}$$

The Magnitude and Direction of the Gradient

Regardless of whether the **Prewitt** or **Sobel** filter is used, the outputs of the filters applied in the horizontal and vertical directions can be interpreted as the **components of the image gradient vector** at each pixel (see Eq. 6.12). These components form the basis for calculating two key edge-related quantities: **gradient magnitude** and **gradient direction** [4].

$$D_x(x, y) = H_x * I \quad \text{and} \quad D_y(x, y) = H_y * I. \tag{6.12}$$

The magnitude of the gradient $E(x, y)$, often denoted as the magnitude of the edge, provides a scalar measure of the strength of intensity change at a pixel location (x, y). This quantity corresponds directly to the edge strength, which is high in regions where intensity varies sharply (i.e., where edges are present), and low in uniform or smoothly varying regions. This magnitude is computed as shown in Eq. 6.13.

$$E(x, y) = \sqrt{(D_x(x, y))^2 + (D_y(x, y))^2}. \tag{6.13}$$

The gradient direction defines the angle at which the intensity change occurs and is perpendicular to the orientation of the edge itself. It is calculated using the arctangent of the ratio between the x and y gradient components, as shown in Eq. 6.14.

$$\phi(x, y) = \tan^{-1}\left(\frac{D_y(x, y)}{D_x(x, y)}\right). \tag{6.14}$$

Figure 6.5 provides a visual representation of both the gradient magnitude and direction on the image, enabling an intuitive understanding of how edges are distributed and oriented.

Figure 6.6 summarizes the complete edge detection process using gradient-based operators. The pipeline includes applying the horizontal H_x and vertical H_y filters,

Fig. 6.5 Illustration of the magnitude $E(x, y)$ and direction $\phi(x, y)$ of the gradient

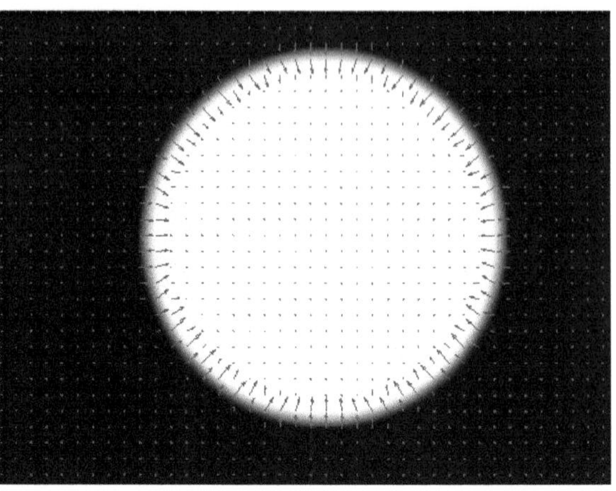

(either Prewitt or Sobel), computing the gradient magnitude $E(x, y)$, and estimating the gradient direction $\phi(x, y)$.

While the **Prewitt** and original **Sobel** operators are both commonly used for edge detection, their accuracy in estimating the **gradient direction** is relatively limited. To address this issue, a **modified version of the Sobel operator** is suggested to **minimize the angular error** in gradient estimation, as shown in Eq. 6.15. This improved filter adjusts the weight distribution within the 3×3 convolution kernel to better capture directional variations in intensity and produce a more accurate representation of the edge orientation.

$$H_x^{S'} = \frac{1}{32}\begin{bmatrix} -3 & 0 & 3 \\ -10 & 0 & 10 \\ -3 & 0 & 3 \end{bmatrix} \quad \text{and} \quad H_y^{S'} = \frac{1}{32}\begin{bmatrix} -3 & -10 & -3 \\ 0 & 0 & 0 \\ 3 & 10 & 3 \end{bmatrix}. \tag{6.15}$$

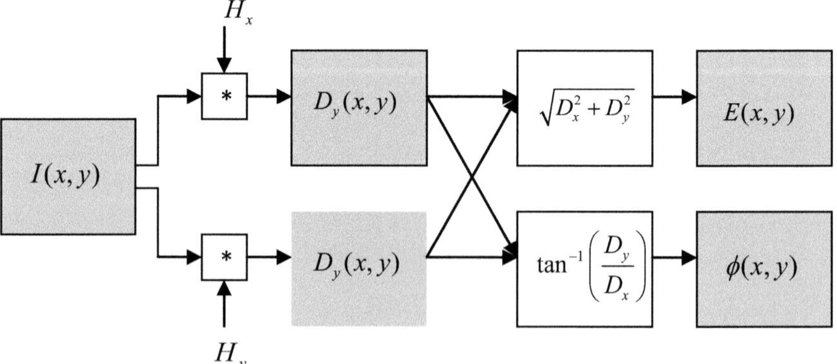

Fig. 6.6 Flow chart of the edge filter process, where H_x and H_y are the kernels, D_x and D_y are the gradients, $E(x, y)$ is the gradient's magnitude, and $\phi(x, y)$ its direction

Despite this limitation, the original Sobel operator remains one of the most widely adopted methods for edge detection. Its popularity is due to a combination of factors, including simplicity of implementation, low computational cost, and the built-in noise reduction provided by the smoothing component of the filter.

As a result, the Sobel operator is integrated into the majority of commercial and open-source digital image processing software libraries, such as OpenCV, MATLAB, and scikit-image, where it is often used as a baseline or preprocessing step for more advanced vision tasks.

6.3.2 The Roberts' Filter

The Roberts operator is recognized as one of the earliest algorithms developed for edge detection in digital images. Introduced in the 1960s, it holds historical significance due to its simplicity and pioneering role in the field of computer vision. Although more advanced methods have since been developed, the Roberts operator remains conceptually essential and continues to be studied as a foundational edge detector.

Technically, the Roberts operator is defined by two 2×2 convolution kernels, which makes it exceptionally compact compared to later edge detectors, such as the Prewitt or Sobel operators. These kernels are designed to estimate the gradient along the diagonals of the image, rather than the vertical or horizontal axes. Specifically, the operator applies the convolution matrices shown in Eq. 6.16.

$$H_1^R = \begin{bmatrix} 0 & 1 \\ -1 & 0 \end{bmatrix} \quad \text{and} \quad H_2^R = \begin{bmatrix} -1 & 0 \\ 0 & 1 \end{bmatrix}. \tag{6.16}$$

Each kernel computes the difference between diagonally adjacent pixels, thus highlighting intensity changes along $45°$ and $135°$ directions. This makes the Roberts operator particularly sensitive to diagonal edges, as illustrated in Fig. 6.7. However, this same characteristic restricts its directional selectivity when applied to edges with similar orientations or when detecting vertical and horizontal boundaries.

The gradient magnitude is derived by combining the responses of both kernels H_1^R and H_2^R can be calculated as described in Eq. 6.5.

Due to the diagonal nature of the convolution, the gradient magnitude can be interpreted geometrically as the resultant vector of two orthogonal components oriented at $45°$, as schematized in Fig. 6.8. This interpretation provides insight into how the filter combines directional changes in intensity to approximate the local gradient at each pixel.

Despite its historical value and minimal computational cost, the Roberts operator is less robust to noise and less accurate in estimating the gradient direction compared to modern filters. As a result, its use in current practical applications is limited.

Fig. 6.7 Illustration of the diagonal kernels of Roberts's filter

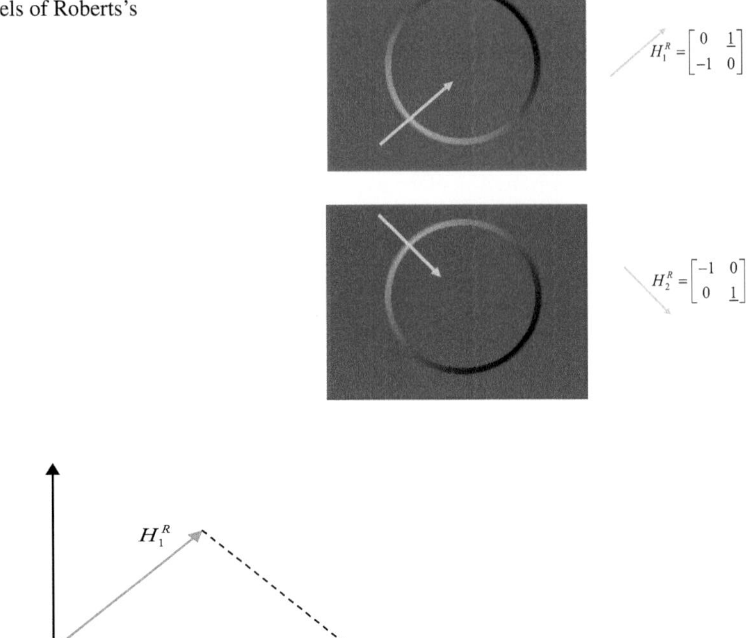

$$H_1^R = \begin{bmatrix} 0 & 1 \\ -1 & 0 \end{bmatrix}$$

$$H_2^R = \begin{bmatrix} -1 & 0 \\ 0 & 1 \end{bmatrix}$$

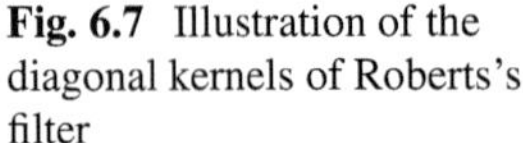

Fig. 6.8 Calculation of the gradient's magnitude $E(x, y)$ in Roberts' method

However, it remains a valuable educational tool for understanding the evolution of edge detection techniques.

6.3.3 Compass Operators

A fundamental challenge in designing edge detection filters lies in balancing two competing objectives: achieving a high response magnitude at edges and maintaining robustness across different edge orientations. Specifically, as filters become more sensitive to intensity transitions, enhancing their ability to detect thin structural boundaries, they also tend to become increasingly direction dependent. This results in diminished detection performance when edges are not aligned with the filter's preferred orientation.

To design an effective edge detector, it becomes essential to strike a compromise between response amplitude and directional sensitivity. Conventional operators like Sobel and Prewitt address this by applying two filters (typically in the x and y axes) while the Roberts operator performs detection along diagonal directions. However, this limited directional sampling can lead to incomplete edge characterization, especially in complex images where structures appear in arbitrary orientations.

One solution to this limitation is to increase the number of directional filters, allowing the operator to detect edges more accurately regardless of their orientation. A classic and influential example of this approach is the Kirsch operator, which employs a set of eight masks, each rotated at $45°$ intervals, thus covering the full $360°$ range of possible edge directions.

Each Kirsch mask is a 3×3 matrix with a high weight in one direction and lower weights in the remaining surrounding positions. These masks are rotated versions of a base filter and are defined in Eq. 6.17.

$$H_0^K = \begin{bmatrix} -1 & 0 & 1 \\ -2 & 0 & 2 \\ -1 & 0 & 1 \end{bmatrix}, H_1^K = \begin{bmatrix} -2 & -1 & 0 \\ -1 & 0 & 1 \\ 0 & 1 & 2 \end{bmatrix},$$

$$H_2^K = \begin{bmatrix} -1 & -2 & -1 \\ 0 & 0 & 0 \\ -1 & 0 & 1 \end{bmatrix}, H_3^K = \begin{bmatrix} 0 & -1 & -2 \\ 1 & 0 & -1 \\ 2 & 1 & 0 \end{bmatrix}, H_4^K = \begin{bmatrix} 1 & 0 & -1 \\ 2 & 0 & -2 \\ 1 & 0 & -1 \end{bmatrix},$$

$$H_5^K = \begin{bmatrix} 2 & 1 & 0 \\ 1 & 0 & -1 \\ 0 & -1 & -2 \end{bmatrix}, H_6^K = \begin{bmatrix} 1 & 2 & 1 \\ 0 & 0 & 0 \\ -1 & -2 & -1 \end{bmatrix}, H_7^K = \begin{bmatrix} 0 & 1 & 2 \\ -1 & 0 & 1 \\ -2 & -1 & 0 \end{bmatrix} \qquad (6.17)$$

From the complete set of eight Kirsch directional filters, denoted as H_0^K, H_1^K, ... H_7^K, only **four unique convolutions** are required for edge detection. This optimization is possible because the last four filters are **mathematically equivalent to the first four**, differing only in sign. For instance, $H_4^K = -H_0^K$. This equivalence is possible due to the linear property of convolution shown in Eq. 6.18.

$$I * H_4^K = I * -H_0^K = -\left(I * H_0^K \right). \qquad (6.18)$$

To apply the Kirsch operator, the filters are convolved with the image as described in Eq. 6.19, and the maximum response at each pixel across all filters is retained. This process not only highlights edges but also determines their dominant orientation, making the Kirsch operator a valuable tool in applications requiring directional edge analysis.

$$D_0 = I * H_0^K \quad D_1 = I * H_1^K \quad D_2 = I * H_2^K \quad D_3 = I * H_3^K,$$

$$D_4 = -D_0 \quad D_5 = -D_1 \quad D_6 = -D_2 \quad D_7 = -D_3. \qquad (6.19)$$

The Kirsch operator defines the magnitude of the gradient as the **maximum response** among the eight directional convolutions (see Eq. 6.20).

$$E^K(x, y) = \max(D_0(x, y), D_1(x, y), \ldots, D_7(x, y))$$
$$= \max(|D_0(x, y)|, |D_1(x, y)|, \ldots, |D_3(x, y)|), \qquad (6.20)$$

The gradient's direction is determined by the filter that provides the maximum response in the gradient magnitude. Thus, the direction of the gradient is specified by Eq. 6.21.

$$\phi^K(x, y) = \frac{\pi}{4} l \quad \text{where} \quad l = \arg\max_{0 \leq i \leq 7}(D_i(x, y)). \qquad (6.21)$$

Although compass operators, such as the Kirsch filter, are theoretically capable of capturing edges in multiple directions with enhanced selectivity, their practical performance does not always offer substantial improvements over simpler methods, like the Sobel operator, in real-world applications. This is particularly true in noisy or low-contrast images, where the complexity of the Kirsch filters may not translate into visibly better edge maps.

Nevertheless, the Kirsch operator offers **computational advantages**, such as avoiding square root calculations for gradient magnitude, eliminating the need for trigonometric functions (e.g., arctangent) for direction estimation, providing robust direction selectivity due to multiple directional filters, and facilitating efficient implementation using only four convolutions, thanks to filter symmetry.

These characteristics make the Kirsch operator especially attractive in **low-resource systems**, embedded platforms, or real-time image processing tasks where **processing speed** and **simplicity** are critical.

6.4   Edge Detection with Python

Building on the theoretical foundation established in the previous sections, we will present how to utilize python programming language tools to calculate an image's edges.

6.4.1 Using Python as a Programming Language to Find Edges

Edge Detection

#CODE 6.1

```python
#The libraries necessary for image processing are loaded.
Import cv2 as cv
import matplotlib.pyplot as plt
import numpy as np
from google.colab.patches import cv2_imshow

#Load image for processing
img = cv.imread('Image6.1.jpg')
img = cv.resize(img,dsize=None,fx=0.2,fy=0.2)
cv2_imshow(img)
```

```python
#The image is transformed from color to grayscale
imgGray = cv.cvtColor(img,cv.COLOR_BGR2GRAY)
cv2_imshow(imgGray)
```

```python
GRADIENT EDGE DETECTION
#One of the 2 filters is applied to the x-axis

#For Prewitt filter
#Hx = np.array([[-1, 0, 1], [-1, 0, 1], [-1, 0, 1]])

#For Sobel filter
#Hx = np.array([[-1, 0, 1], [-2, 0, 2], [-1, 0, 1]])

# For this example we will use the Sobel filter
Hx = np.array([[-1, 0, 1], [-2, 0, 2], [-1, 0, 1]])
Gx = cv.filter2D(np.double(imgGray),-1,Hx)
Gx_img = Gx.copy()
min = np.min(Gx_img)
Gx_img = Gx_img - min
max = np.max(Gx_img)
Gx_img = (Gx_img / max) * 255
cv2_imshow(np.uint8(Gx_img))
```

```python
#Another alternative in Python is to use the
Sklearn method to avoid calculating minimums and maximums
from sklearn.preprocessing import MinMaxScaler

#And we apply the same filter as above to the x-axis

#For Prewitt filter
#Hx = np.array([[-1, 0, 1], [-1, 0, 1], [-1, 0, 1]])

#Sobel filter for this case
Hx = np.array([[-1, 0, 1], [-2, 0, 2], [-1, 0, 1]])

Gx = cv.filter2D(np.double(imgGray),-1,Hx)
scaler = MinMaxScaler(feature_range=(0, 255))
```

```python
#Before scaling, it is convenient to convert to vector and then
to matrix since the scaling is done by column.
Gx_img = scaler.fit_transform(Gx.reshape(-1,1)).reshape(Gx.shape)
cv2_imshow(np.uint8(Gx_img))
```

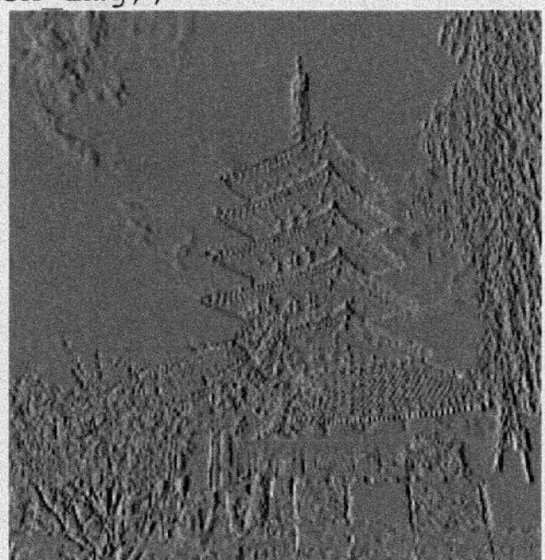

```python
# The filter is applied to the y-axis

#For Prewitt filter
#Hy = np.array([[-1, -1, -1], [0, 0, 0], [1, 1, 1]])

#For Sobel filter
#Hy = np.array([[-1, -2, -1], [0, 0, 0], [1, 2, 1]])
# For this example we will use the Sobel filter
Hy = np.array([[-1, -2, -1], [0, 0, 0], [1, 2, 1]])
Gy = cv.filter2D(np.double(imgGray),-1,Hy)
Gy_img = Gy.copy()
min = np.min(Gy_img)
Gy_img = Gy_img - min
max = np.max(Gy_img)
Gy_img = (Gy_img / max) * 255
cv2_imshow(np.uint8(Gy_img))
```

```
#Filter is applied to the y-axis but using Sklearn method
Gy = cv.filter2D(np.double(imgGray),-1,Hy)
Gy_img = scaler.fit_transform(Gy.reshape(-1,1)).reshape(Gy.shape)
cv2_imshow(np.uint8(Gy_img))
```

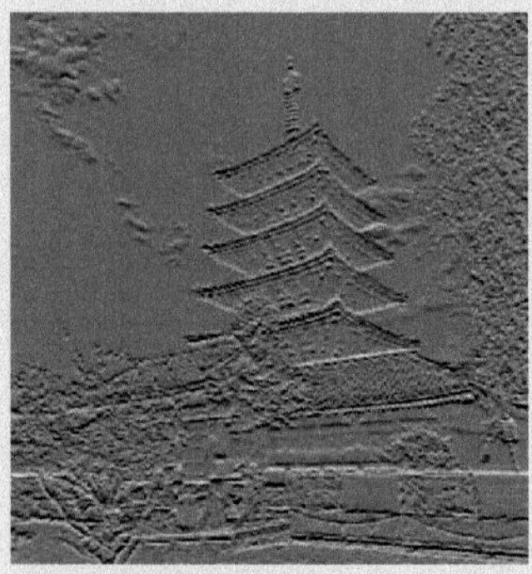

```
#Total gradient is calculated and normalized by dividing by the
largest value and multiplying by 255.
Gtotal = np.sqrt(Gx**2 + Gy**2)
max = np.max(Gtotal)
print(max)
Gtotal = np.uint8((Gtotal / max) * 255)
cv2_imshow(Gtotal)
```

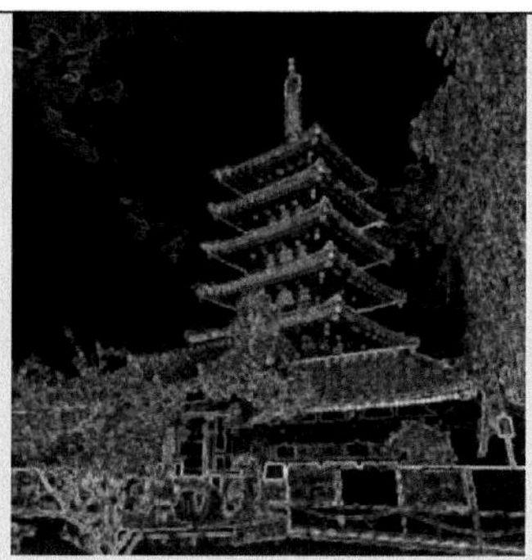

```
#Total gradient is calculated and normalized but using the
sklearn method
Gtotal = np.sqrt(Gx**2 + Gy**2)

Gtotal = scaler.fit_transform(Gtotal.reshape(-
1,1)).reshape(Gtotal.shape)

cv2_imshow(np.uint8(Gtotal))
```

```
#The histogram is drawn to select a threshold.
plt.figure()
plt.hist(Gtotal.flatten(),bins=256)
plt.show()
```

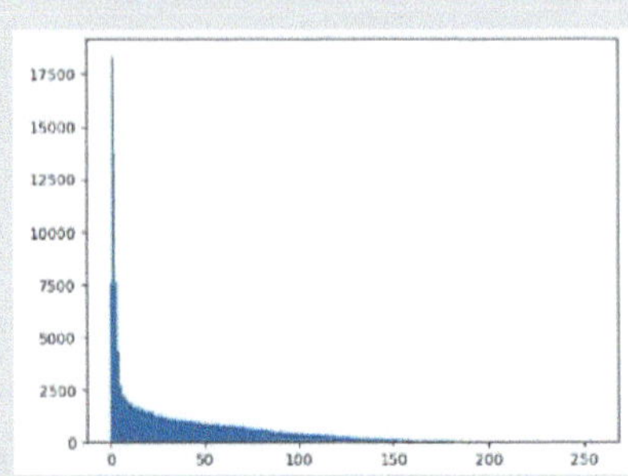

```
#Binarizes with the selected threshold
# The following commands return the binarized image and the
threshold used by the thresh method

_, GtotalBin =
cv.threshold(Gtotal,thresh=40,maxval=255,type=cv.THRESH_BINARY
cv2_imshow(GtotalBin)
```

Edge Detection with Sobel Filter

#CODE 6.2

```python
imgGrayGauss = cv.GaussianBlur(imgGray,(3,3),0)
cv2_imshow(imgGrayGauss)
```

```python
from sklearn.preprocessing import MinMaxScaler

sobel_x =
cv.Sobel(imgGrayGauss,ddepth=cv.CV_64F,dx=1,dy=0,ksize=3)

scaler = MinMaxScaler(feature_range=(0, 255))
sobel_x_img = scaler.fit_transform(sobel_x.reshape(-
1,1)).reshape(sobel_x.shape)

cv2_imshow(np.uint8(sobel_x_img))
```

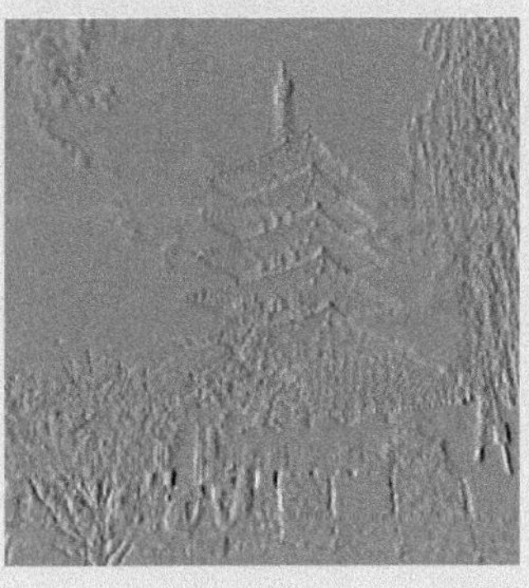

```
sobel_y =
cv.Sobel(imgGrayGauss,ddepth=cv.CV_64F,dx=0,dy=1,ksize=3)
sobel_y_img = scaler.fit_transform(sobel_y.reshape(-
1,1)).reshape(sobel_y.shape)
cv2_imshow(np.uint8(sobel_y_img))
```

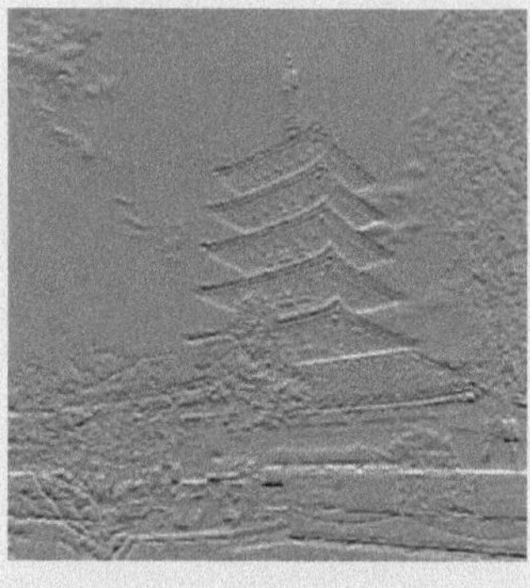

```
imgSobel = np.sqrt(sobel_x**2 + sobel_y**2)
max = np.max(imgSobel)
print(max)
imgSobel = np.uint8((imgSobel / max) * 255)
cv2_imshow(imgSobel)
```

```
#Binarizes with the selected threshold
_, imgSobelBin = cv.threshold(imgSobel,40,255,cv.THRESH_BINARY)
cv2_imshow(imgSobelBin)
```

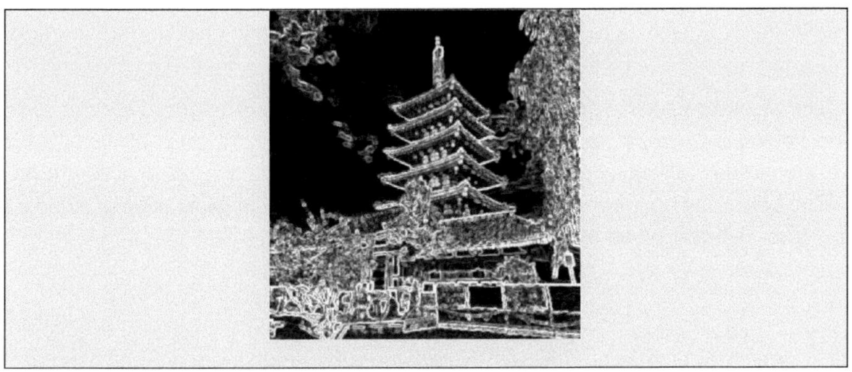

6.5 Second-Derivative Edge Filters

As previously discussed in Sect. 6.2, edge detection methods based on the first derivative rely on estimating the image gradient. These methods identify edges as regions with high gradient magnitudes, i.e., areas of abrupt intensity change in specific directions.

However, an alternative class of border detection operators uses the second derivative of the image function. These methods identify edges not by the strength of the gradient, but by detecting zero-crossings of the second derivative, which means a change of sign due to a negative or positive gradient change.

While detecting the presence of an edge is fundamental, certain applications, such as object detection and feature analysis, may also require distinguishing between light-to-dark and dark-to-light transitions. Therefore, the second derivative-based methods are helpful because the zero-crossing reveals whether the pixel is on the light or dark side of the edge.

6.5.1 Second-Derivative Method for Edge Detection

The second-derivative method consists of finding the zero-crossing, which is the transition from negative to positive or positive to negative. This change is found when calculating the second derivative.

For a one-dimensional function, the definition of the first derivative is stated in Eq. 6.22.

$$\frac{\partial f}{\partial x} = f(x+1) - f(x). \tag{6.22}$$

The expression of the second derivative shown in Eq. 6.23 can be formulated from Eq. 6.22.

$$\frac{\partial f^2}{\partial x^2} = f(x+1) - 2f(x) + f(x-1). \tag{6.23}$$

The second derivative provides the gradient value in any direction. This characteristic, known as isotropic, makes it insensitive to rotation, meaning it is less sensitive to the orientation of the edge and thus may offer better localization in certain conditions.

The Laplacian operator is an isotropic filter defined by the expression of Eq. 6.24. This filter is linear when using derivatives [5].

$$\nabla^2 I(x, y) = \frac{\partial^2 I(x, y)}{\partial x^2} + \frac{\partial^2 I(x, y)}{\partial y^2}. \tag{6.24}$$

Considering Eq. 6.23 in the Laplacian expression of Eq. 6.24, we obtain Eqs. 6.25 and 6.26.

$$\frac{\partial^2 I(x, y)}{\partial x^2} = I(x+1, y) - 2I(x, y) + I(x-1, y), \tag{6.25}$$

$$\frac{\partial^2 I(x, y)}{\partial y^2} = I(x, y+1) - 2I(x, y) + I(x, y-1). \tag{6.26}$$

Then, if we substitute Eqs. 6.25 and 6.26 in 6.24, we can formulate Eq. 6.27.

$$\nabla^2 I(x, y) = I(x+1, y) + I(x-1, y) + I(x, y+1) + I(x, y-1) - 4I(x, y). \tag{6.27}$$

If the above expression were represented as a convolutional filter that could be used for calculating the second derivative, it would be defined as displayed in Fig. 6.9 (Fig. 6.10).

An illustrative example of the effect on a grayscale image when applying the Laplacian filter is presented in Fig. 6.11. This figure highlights the outcomes of using the second-order derivatives separately along the horizontal and vertical directions, followed by their combination, which forms the classical Laplacian operator. The Laplacian, as defined in Eq. 6.24 and implemented through the filter shown in Fig. 6.9, initially considers only the four immediate (orthogonal) neighbors of a pixel (top, bottom, left, and right), thus excluding the influence of diagonal neighbors.

Fig. 6.9 Laplacian filter obtained from Eq. 6.27

0	1	0
1	-4	1
0	1	0

<table>
<tr><td>1</td><td>1</td><td>1</td></tr>
<tr><td>1</td><td>-8</td><td>1</td></tr>
<tr><td>1</td><td>1</td><td>1</td></tr>
</table>

Fig. 6.10 Laplacian filter, considering diagonal neighbors

(a)   (b)

(c)   (d)

Fig. 6.11 Example of the Laplacian operator for edge detection. **a** Grayscale image, **b** derivative $\partial^2 I(x, y)/\partial x^2$, **c** derivative $\partial^2 I(x, y)/\partial y^2$, and **d** the Laplacian filter $\nabla^2 I(x, y)$

To account for these diagonal variations, the filter can be extended by incorporating the diagonals into the kernel. This modification adds four additional ones to the kernel's coefficient matrix. To maintain the zero-sum property characteristic of Laplacian filters, essential for preserving their high-pass behavior, it becomes necessary to increase the central coefficient from 4 to 8. The resulting modified Laplacian kernel is presented in Fig. 6.10.

This enhancement enables the operator to capture isotropic changes in intensity more effectively, making it more responsive to edge variations that do not align strictly with horizontal or vertical orientations. It also improves edge localization in complex image structures.

A key characteristic of Laplacian filters is that they yield a response of zero in regions of uniform intensity (i.e., constant grayscale areas), and nonzero responses where intensity transitions occur. This is due to the filter's inherent nature as a second-order derivative operator, which is sensitive to changes in the first derivative, thereby identifying areas where the rate of intensity change is significant.

Furthermore, because the Laplacian operator acts as a high-pass filter (as indicated by the zero sum of its kernel coefficients), it effectively enhances fine details and sharp transitions in the image while suppressing low-frequency components such as smooth background regions.

6.6   Improved Image Sharpness

When the Laplacian filter is applied to an image, it primarily highlights regions of rapid intensity change, effectively revealing the edges within the image. However, if the objective is to enhance the sharpness of the image rather than merely extract its edges, a different approach must be taken. In this case, it is essential to preserve the image's low-frequency content (corresponding to the general structure and smooth intensity variations) while simultaneously enhancing its high-frequency components, which correspond to fine details and edges.

This enhancement can be achieved by combining the original image with a scaled version of its Laplacian-filtered counterpart. Specifically, instead of directly using the Laplacian-filtered image as the final result, one subtracts a proportion of the Laplacian response from the original image. This subtraction amplifies the high-frequency details (i.e., edges and tinny structures) without discarding the base intensity infor-

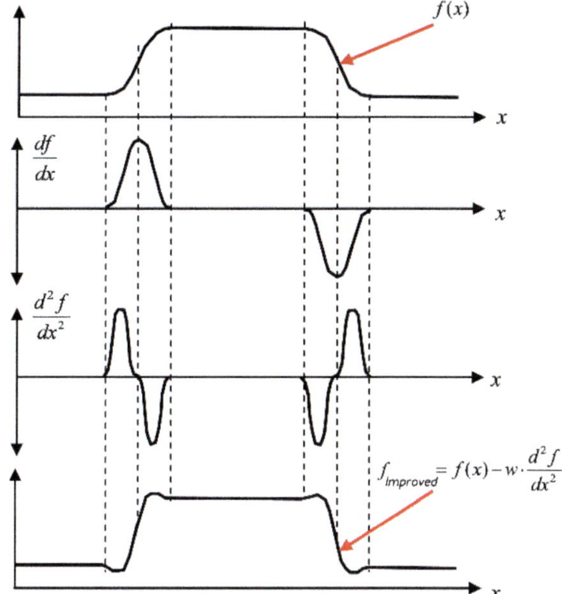

Fig. 6.12 Increased sharpness through the application of the second derivative. The contours in the image can be maximized if a weighted second derivative is subtracted from the image function

mation. The result is an image with improved sharpness, as the edges become more pronounced while maintaining a natural overall appearance. Equation 6.28 describes how to improve sharpness using the Laplacian filter.

$$I(x, y)_{\text{Improved}} = I(x, y) - w \cdot \nabla^2 I(x, y). \tag{6.28}$$

Figure 6.12 illustrates the idea whereby the sharpness of the image is enhanced when its edges are more evident. For ease of explanation, only the one-dimensional case is considered.

The enhancement of the image sharpness can be achieved in one step if the scale $w = 1$ as shown in Eq. 6.29. Thus, considering the Laplacian described in Eq. 6.30, we can generate the enhanced image in a single step by using Eq. 6.31. This convolutional

(a)   (b)

Fig. 6.13 Example of sharpening an image using the Laplacian operator. **a** Grayscale image, and **b** the result of applying $I(x, y)_{\text{Improved}} = I(x, y) - w \cdot \nabla^2 I(x, y)$, with $w = 1$

filter is shown in Eq. 6.32.

$$I(x, y)_{\text{Improved}} = I(x, y) - (1) \cdot \nabla^2 I(x, y), \tag{6.29}$$

$$\nabla^2 I(x, y) = I(x + 1, y) + I(x - 1, y) + I(x, y + 1) + I(x, y - 1) - 4I(x, y), \tag{6.30}$$

$$I(x, y)_{\text{Improved}} = 5I(x, y) - \big[I(x + 1, y) + I(x - 1, y) \\ + I(x, y + 1) + I(x, y - 1)\big], \tag{6.31}$$

$$I(x, y)_{\text{Improved}} = \begin{bmatrix} 0 & -1 & 0 \\ -1 & 5 & -1 \\ 0 & -1 & -1 \end{bmatrix}. \tag{6.32}$$

Figure 6.13 shows the sharpness enhancement of an image using a $w = 1$.

6.6.1 The Laplacian Filter for Image Sharpness Enhancement in Python

In this section, an example of edge detection using the Laplacian filter is presented. Since the Laplacian filter is very sensitive to noise, a Gaussian filter must be applied first.

Edge Detection with Laplacian Filter

#CODE 6.3

```python
# A Gaussian filter is applied to remove and smooth noise.
imgGray = cv.GaussianBlur(imgGray, (3,3),0)
cv2_imshow(imgGray)
```

```python
#Laplacian filter for the x-axis
laplace_x = np.array([[0, 1, 0], [0, -2, 0], [0, 1, 0]])
Laplace_x = cv.filter2D(np.double(imgGray),-1,laplace_x)
min = np.min(Laplace_x)
imgLaplace_x = Laplace_x - min
max = np.max(imgLaplace_x)
imgLaplace_x = np.uint8((imgLaplace_x / max)*255)
cv2_imshow(imgLaplace_x)
```

```
#Laplacian filter for the y-axis
laplace_y = np.array([[0, 0, 0], [1, -2, 1], [0, 0, 0]])
Laplace_y = cv.filter2D(np.double(imgGray),-1,laplace_y)
min = np.min(Laplace_y)
imgLaplace_y = Laplace_y - min
max = np.max(imgLaplace_y)
imgLaplace_y = np.uint8((imgLaplace_y / max) * 255)
cv2_imshow(imgLaplace_y)
```

```
#Total Laplacian
imgLaplaceTotal = Laplace_x + Laplace_y
cv2_imshow(np.uint8(imgLaplaceTotal))
```

```
#Laplacian with full kernel (two dimensions in one kernel)
kernel_laplace = np.array([[0, 1, 0], [1, -4, 1], [0, 1, 0]])
imgLaplace = cv.filter2D(np.double(imgGray),-1,kernel_laplace)
cv2_imshow(np.uint8(imgLaplace))
```

```python
#Laplacian with full kernel considering diagonal neighbors
kernel_laplace = np.array([[1, 1, 1], [1, -8, 1], [1, 1, 1]])
imgLaplace = cv.filter2D(np.double(imgGray),-1,kernel_laplace)
cv2_imshow(np.uint8(imgLaplace))
```

```python
#Laplacian with function cv.Laplacian()
# ddepth is the data type
imgLaplacian = cv.Laplacian(imgGray,ddepth=cv.CV_64F)
cv2_imshow(np.uint8(imgLaplacian))
```

```python
#The image is transformed from color to grayscale
img2Gray = cv.cvtColor(img2,cv.COLOR_BGR2GRAY)
cv2_imshow(img2Gray)
```

```python
#One of the 2 Laplacian filters are applied
```

```
Sharpness Improvement with Laplacian Filter

        #CODE 6.4


#The libraries necessary for image processing are loaded.
import cv2 as cv
import matplotlib.pyplot as plt
import numpy as np
from google.colab.patches import cv2_imshow

#Load the image for processing
img = cv.imread('Image6.2.jpg')
img2 = cv.resize(img,dsize=None,fx=0.5,fy=0.5)
cv2_imshow(img2)
```

```
#Space Laplacian filter of an image
kernel_laplace = np.array([[0, 1, 0], [1, -4, 1], [0, 1, 0]])

#Laplacian filter with the extension of diagonal neighbors to the
central pixel
#kernel_laplace = np.array([[1, 1, 1], [1, -8, 1], [1, 1, 1]])

# For this example, we will use the spatial Laplacian filter of
an image
kernel_laplace = np.array([[0, 1, 0], [1, -4, 1], [0, 1, 0]])
img2Laplace = cv.filter2D(np.double(img2Gray),-1,kernel_laplace)
imgNitida = img2Gray - img2Laplace
cv2_imshow(imgNitida)
```

```
imgLaplacian = cv.Laplacian(img2Gray,ddepth=cv.CV_64F)
imgNitida = img2Gray - imgLaplacian
cv2_imshow(imgNitida)
```

6.7   Canny Filter

A recognized technique for edge detection is the Canny filter. This edge detection scheme considers applying a set of kernel filters in various directions and with different resolutions, which, when combined, extract the borders from the image. The Canny algorithm tries to reach three particular objectives: (1) reduce the number of false edges, (2) better localization of the border in the image, and (3) generate an image with one-pixel edge width. The Canny filter, in essence, is based on gradient calculations. Still, it is also used for edge localization as a criterion for the second derivative or Laplacian. This method is usually used in its simple form, i.e., by configuring just the smoothing parameter σ. Figure 6.14 displays the result of applying the Canny algorithm with different values of sigma.

One of the most widely recognized and effective methods for edge detection in digital images is the **Canny filter**, proposed by John F. Canny. This method is designed to identify optimal edges by combining gradient-based detection with a

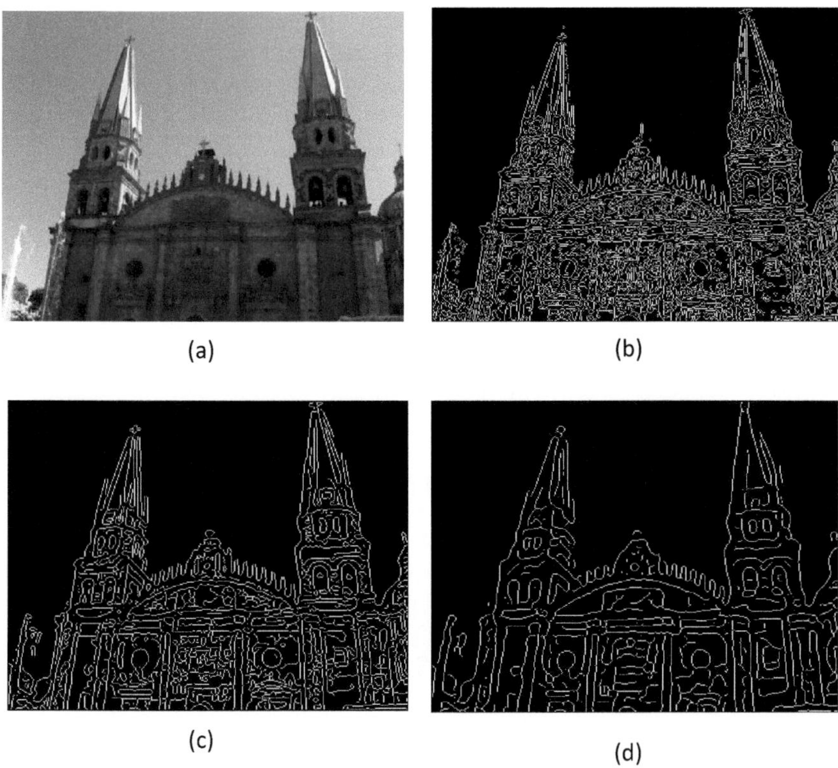

Fig. 6.14 Application of the Canny method. **a** Grayscale image, **b** edges considering $\sigma = 0.1$, **c** edges considering $\sigma = 1$, and **d** edges with $\sigma = 2$

multistage processing approach. The Canny method aims to satisfy three essential criteria: (a) **low error rate**, meaning it minimizes the number of false positives (detecting non-existent edges) and false negatives (missing actual edges); (b) **effective localization**, ensuring that the detected edges are as close as possible to the actual boundaries within the image; and (c) accurate detection, aiming to produce edges that are as thin as possible, ideally one pixel wide, to maintain precision in the representation.

The Canny filter operates through a multi-step process. First, the image is **smoothed with a Gaussian filter**, which reduces noise and small fluctuations in intensity that might lead to spurious edge detection. The smoothing level is controlled by the σ **(sigma)** parameter; a higher σ value results in greater smoothing, which can suppress weak edges but enhance robustness to noise. After smoothing, the **gradient magnitude and direction** are computed, typically using first-derivative operators, such as the Sobel operator.

Next, the method applies **non-maximum suppression**, which thins the border by considering just the local maximum. This is followed by applying two thresholds, which identify borders as strong, weak, or irrelevant. Finally, **border tracking by hysteresis** is performed to connect weak edge pixels to strong ones if they are part of the same edge contour, effectively reducing isolated responses and reinforcing edge continuity.

Although fundamentally based on first-derivative methods, the Canny filter's consideration of gradient direction and the application of non-maximum suppression introduces an indirect dependence on second-derivative concepts (similar to the Laplacian). The method is typically used by configuring only the smoothing parameter σ, which determines the scale of edge detection.

Figure 6.14 illustrates how different values of σ influence the results: lower values capture finer details but may increase sensitivity to noise. In contrast, higher values produce smoother edge maps, often at the cost of omitting subtle structures.

6.7.1 Implementing the Canny Filter with Python

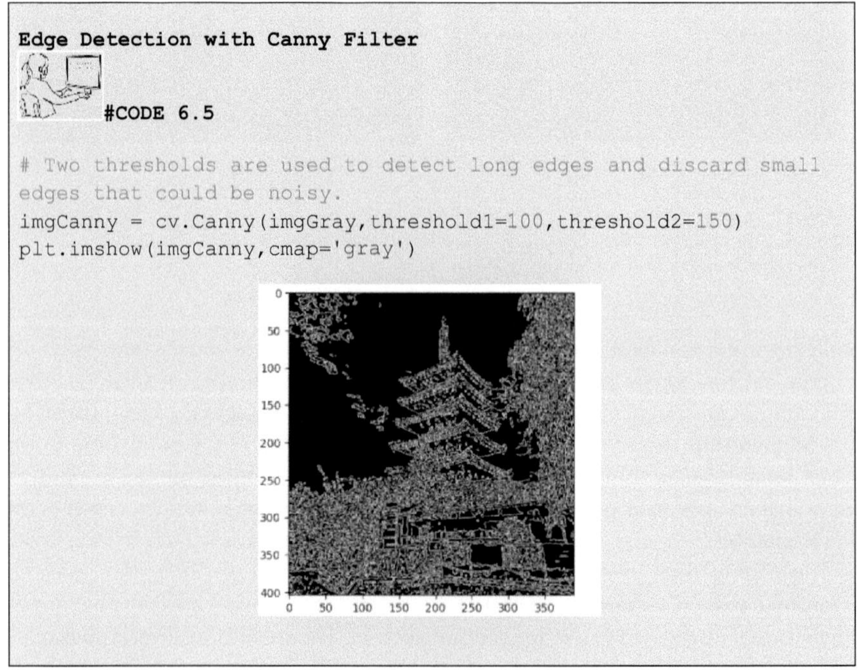


```python
# Two thresholds are used to detect long edges and discard small
edges that could be noisy.
imgCanny = cv.Canny(imgGray,threshold1=100,threshold2=150)
plt.imshow(imgCanny,cmap='gray')
```

References

1. Canny, J. F. (1986). A computational approach to edge detection. *IEEE Transactions on Pattern Analysis and Machine Intelligence, 8(6)*, 679–698.
2. Gonzalez, R. C., & Woods, R. E. (2018). *Digital image processing* (4th ed.). Pearson.

3. Marr, D., & Hildreth, E. (1980). Theory of edge detection. *Proceedings of the Royal Society of London. Series B, Biological Sciences, 207*(1167), 187–217.
4. Sonka, M., Hlavac, V., & Boyle, R. (2014). *Image processing, analysis, and machine vision* (4th ed.). Cengage Learning.
5. Pratt, W. K. (2007). *Digital image processing: PIKS inside* (4th ed.). Wiley-Interscience.

Chapter 7
Determination of Corners

This chapter presents an study of corner detection as a key process in image analysis and computer vision. It begins by defining corners as points of high gradient variation in multiple directions, typically located where prominent edges intersect. The chapter emphasizes their importance in tasks such as object recognition, motion tracking, 3D reconstruction, and camera calibration, due to their stability under illumination and perspective changes. The discussion centers on the Harris–Stephens corner detection algorithm, explaining its mathematical foundation through the computation of image gradients, the construction of the structure (or second-moment) matrix, and the use of eigenvalue analysis to distinguish corners from edges and flat regions. The derivation of the corner response function $V(x, y) = \det(E) - \alpha(\mathrm{tr}(E))^2$ is detailed, highlighting how the parameter α controls detection sensitivity. The algorithm's implementation is systematically outlined—from image smoothing and gradient computation to thresholding and local maximum selection for precise corner localization. In addition, the chapter provides Python implementations using both manual coding and built-in functions from OpenCV and scikit-image, demonstrating practical methods for detecting and visualizing corners in real images. Through theoretical derivation, algorithmic steps, and applied coding examples, this chapter offers a comprehensive framework for understanding and implementing robust corner detection methods in digital image processing.

The edges discussed in the previous chapter were defined as points with a high gradient value. Simultaneously, corners represent significant features within an image due to their high gradient magnitude. Unlike edges, where the gradient changes mainly in one direction, corners exhibit notable variation across multiple directions. Based on this idea, corners can be visualized as specific points where distinct edges intersect or converge within the image space.

Corners are applicable in numerous tasks, including object tracking in video streams, defining structural layouts in stereo imaging, serving as reference markers in geometric analysis of shapes, and assisting in the calibration of visual sensors in imaging systems. A key benefit of corners, compared to other image-based features,

© The Author(s), under exclusive license to Springer Nature Switzerland AG 2026

E. Cuevas et al., *Image Processing with Python*, Signals and Communication Technology, https://doi.org/10.1007/978-3-032-13285-7_7

lies in their stability under perspective transformations and their consistent detection even when lighting conditions vary.

7.1   Intersection Points of Prominent Edges

Given that corners represent strong and noticeable attributes within an image, identifying them, as will be discussed later, involves a certain level of complexity. A reliable corner detection method should satisfy various critical requirements, including:

- Distinguishing "significant" corners from less relevant ones.
- Functioning effectively even when the image contains inherent noise.
- Quick processing to enable use in real-time applications.

It is reasonable to assume that multiple strategies can meet these requirements. Nonetheless, many of them rely on evaluating the gradient at points suspected to be corners. While an edge corresponds to a location with a high gradient in a single direction, a corner exhibits significant gradient values across multiple directions at once.

Common corner detection algorithms [1] typically employ either first- or second-order derivatives to estimate or approximate the gradient magnitude in various orientations. One well-known method that follows this principle is the Harris Detector (also known as the Plessey corner detector). Although alternative techniques offer notable capabilities, the Harris approach remains the most widely adopted, and therefore, it will be examined in further detail.

Corner detection methods are widely used in image processing and computer vision for identifying points in an image where the intensity changes sharply in multiple directions—typically indicating features like object corners, junctions, or texture variations. These points are highly informative and are often used as keypoints in tasks such as object recognition, image matching, and motion tracking. For example, in stereo vision and 3D reconstruction, corresponding corners across multiple views help determine depth and spatial relationships. In robotics and navigation, corners serve as reliable landmarks for simultaneous localization and mapping (SLAM). In augmented reality, they enable accurate alignment of virtual objects with the physical environment. Additionally, in image stitching or panorama creation, corner points are used to align overlapping images by finding geometric transformations. Methods such as Harris, Shi-Tomasi, and FAST provide robust corner detection under varying lighting and perspective conditions, making them essential tools in feature-based computer vision pipelines.

7.2 Harris Corner Detection Method

The Harris–Stephens algorithm [2] operates on the principle that corners represent image points exhibiting significant gradient magnitudes across multiple directional orientations simultaneously. This algorithm is robust in differentiating the corners from the edges, which show a high gradient value, but in only one direction; this detector also allows for robustly locating the corners to the orientation, so the orientation of the corner does not matter.

7.2.1 Structure Matrix

The Harris algorithm computes corner responses by evaluating first-order partial derivatives of pixel $I(x, y)$ **along both x and y axes [3], expressed mathematically as:**

$$I_x(x, y) = \frac{\partial I(x, y)}{\partial x} \text{ and } I_y(x, y) = \frac{\partial I(x, y)}{\partial y}. \tag{7.1}$$

Three distinct values are computed for every image pixel (x, y), designated as $HE_{11}(x, y)$, $HE_{22}(x, y)$, and $HE_{12}(x, y)$, where these terms represent:

$$HE_11(x, y) = I_x^2(x, y), \tag{7.2}$$

$$HE_22(x, y) = I_y^2(x, y), \tag{7.3}$$

$$HE_{12}(x, y) = I_x(x, y) \cdot I_y(x, y). \tag{7.4}$$

These parameters serve as numerical estimates for the components of the structural matrix HE, formally expressed as:

$$HE = \begin{bmatrix} HE_{11} & HE_{12} \\ HE_{21} & HE_{22} \end{bmatrix}, \tag{7.5}$$

i.e., $HE_{12} = HE_{21}$.

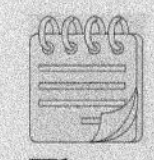

The structure matrix, also known as the second-moment matrix, is widely used beyond corner detection in several areas of computer vision and image processing due to its ability to describe local intensity variations. In optical flow estimation, the structure matrix is essential for tracking motion by analyzing how pixel intensities shift between consecutive frames. It is also used in image registration, where accurate alignment of overlapping images depends on identifying and matching features based on local gradients. In feature extraction and matching, especially in scale- or rotation-invariant algorithms like SIFT, the structure matrix helps determine keypoints with stable local patterns. Furthermore, in texture analysis, it can characterize surface patterns by capturing directional changes in intensity. Because it encodes directional and magnitude information about gradients, the structure matrix is a versatile and fundamental tool for analyzing and interpreting image structure in dynamic or static contexts.

7.2.2 Matrix HE Spectral Filtering

To identify corners through the Harris method [4], each component of the structure matrix must be smoothed by applying a Gaussian filter H_σ. This convolution process adjusts the values, resulting in a redefined matrix E that replaces the original HE matrix according to the following relation:

$$E = \begin{bmatrix} \text{HE}_{11} * H_\sigma & \text{HE}_{12} * H_\sigma \\ \text{HE}_{21} * H_\sigma & \text{HE}_{22} * H_\sigma \end{bmatrix} = \begin{bmatrix} A & C \\ C & B \end{bmatrix}. \tag{7.6}$$

7.2.3 Eigenvalue–Eigenvector Extraction

Thanks to its symmetrical nature, the matrix E can be diagonalized such that:

$$E' = \begin{bmatrix} \lambda_1 & 0 \\ 0 & \lambda_2 \end{bmatrix}, \tag{7.7}$$

where λ_1 and λ_2 represent the eigenvalues of matrix E and are determined based on the following expression:

$$\lambda_{1,2} = \frac{\text{tr}(E)}{2} \pm \sqrt{\left(\frac{\text{tr}(E)}{2}\right)^2 - \det(E)}, \tag{7.8}$$

where $\text{tr}(E)$ refers to the trace of matrix E, and $\det(E)$ corresponds to its determinant. By expanding the operations for both the trace and determinant as given in Eq. 7.7,

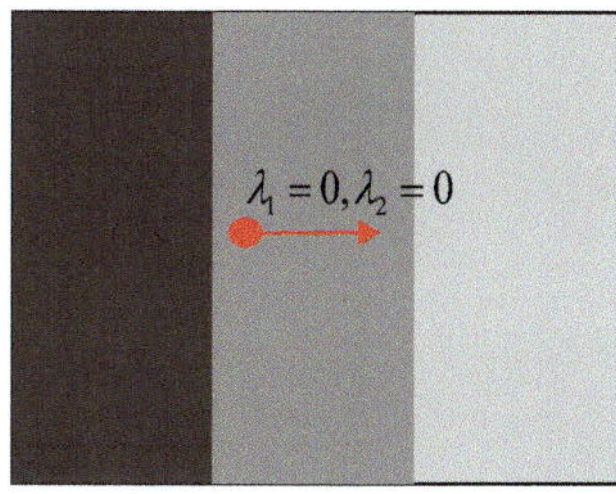

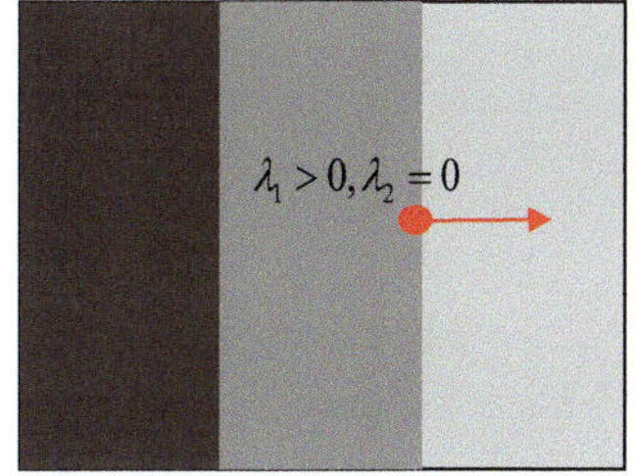

Fig. 7.1 Connection between image structures and their corresponding eigenvalues

we arrive at:

$$\lambda_{1,2} = \frac{1}{2}(A + B \pm \sqrt{A^2 - 2AB + B^2 + 4C^2}). \tag{7.9}$$

The eigenvalues λ_1 and λ_2 are both positive and provide valuable insight into the local patterns present in the image. In an area where the intensity remains constant, resulting in a homogeneous region, the matrix E becomes zero, and thus, both eigenvalues take the value $\lambda_1 = \lambda_2 = 0$. However, the value of $\lambda_1 > 0$ and $\lambda_2 = 0$ in an intensity step is changed, regardless of the edge orientation. Based on the previous explanation, the eigenvalues represent the strength of the gradient, whereas the eigenvectors indicate its orientation. Figure 7.1 illustrates how the values of λ_1 and λ_2 provide key details about the structural elements within an image.

A corner is identified when the gradient magnitude is high in both the main orientation linked to the values of both eigenvalues and the perpendicular direction associated with the smaller eigenvalue. Based on this, it is inferred that a point can be classified as a corner only if both eigenvalues λ_1 and λ_2 have considerable magnitudes. Given that A and C are non-negative, the trace of matrix E, $\mathrm{tr}(E) > 0$. From this and by analyzing Eqs. 7.8 and 7.9, it follows that $|\lambda_1| \geq |\lambda_2|$. Therefore, since both eigenvalues must be significantly large, focusing on λ_2, the smaller of the two, is sufficient for detecting corners: if λ_2 is large, then λ_1 must be as well.

7.2.4 *Corner Response Measure (V)*

Equations 7.8 and 7.9 formally express the eigenvalue divergence as:

$$\lambda_1 - \lambda_2 = 2 \cdot \sqrt{\left(\frac{\mathrm{tr}(E)}{2}\right)^2 - \det(E)}, \tag{7.10}$$

where in any case, it is satisfied that $(1/4) \cdot \mathrm{tr}(E) > \det(E)$. Corner points exhibit minimal eigenvalue divergence. The Harris method exploits this characteristic, defining the corner response function as:

Fig. 7.2 Visualization of V(x, y) values across the image

$$V(x, y) = \det(E) - \alpha(\mathrm{tr}(E))^2, \tag{7.11}$$

which, according to Eq. 7.6, leads to:

$$V(x, y) = \left(A \cdot B - C^2\right) - \alpha(A + B)^2. \tag{7.12}$$

In this context, the parameter α regulates the sensitivity of the algorithm through the function $V(x, y)$, which represents the corner response at a given point. A higher value of $V(x, y)$ indicates a stronger presence of a corner at coordinates (x, y). The constant α is typically chosen within the range $[0.04, 0.25]$. When α is larger, the algorithm becomes less responsive to corners; for instance, selecting $\alpha = 0.04$ increases the algorithm's sensitivity. Consequently, a smaller α results in the detection of more corner points within an image. Figure 7.2 illustrates how $V(x, y)$ behaves across an image, showing that corner points produce high values while flat or edge regions yield values close to zero.

7.2.5 *Corner Point Identification and Localization*

A pixel at coordinates (x, y) qualifies as a candidate corner when meeting the specified criteria.

$$V(x, y) > t_h. \tag{7.13}$$

Here, the threshold value is applied, and it usually falls within the interval of 900 to 10,000, depending on the characteristics of the specific image. By using Eq. 7.13, a binary matrix is generated, where elements are marked with ones (true) if the condition is satisfied, and with zeros (false) if the condition does not hold.

To avoid identifying densely clustered regions of corner points caused by a low α value and resulting high sensitivity, the selection is limited to pixels where the $V(x, y)$ value is the highest within a specified neighborhood area. This filtering step helps isolate the most representative corners. Figure 7.3 provides a visual example of this procedure.

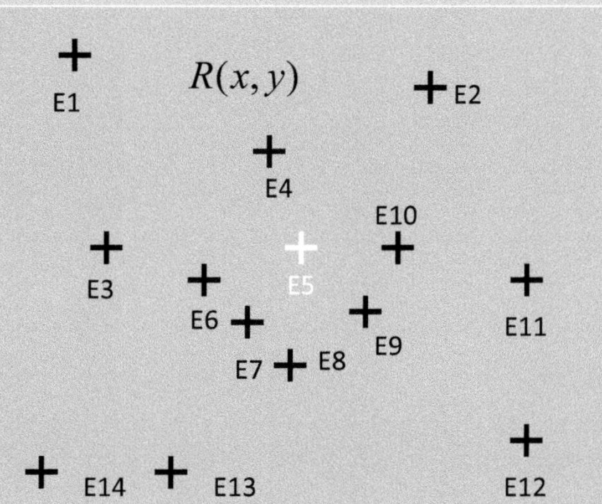

Fig. 7.3 Procedure for selecting the most relevant corner. Among all corner candidates identified using the threshold t_h, the algorithm chooses the point with the highest $V(x, y)$ value within a specified neighborhood region $R(x, y)$. In this example, corner $E5$ stands out, as its $V(x, y)$ value surpasses those of surrounding points $E4$, $E6$, $E7$, $E8$, $E9$, and $E10$ within the defined neighborhood $R(x, y)$

7.2.6 Algorithm Coding and Execution

The Harris algorithm represents the first "advanced" technique introduced in this book; this section offers a comprehensive overview of the method and outlines each stage required to detect corner points within an image. A full summary of the procedure is detailed in the steps provided by Algorithm 7.1.

Algorithm 7.1 Harris algorithm for detection of corner points

	Harris Detector $(I(x, y))$
1:	Smoothing of the original image by a pre-filter: $I' = I * H_p$
2:	Step 1: Compute the corner response function $V(x, y)$
3:	Horizontal (I_x) and vertical (I_λ) gradient components are computed
4:	$I_x = I' * H_x$ y $I_y = I' * H_y$
5:	Structure matrix components $(HE_{11}, HE_{12}, HE_{21}, HE_{22})$ are computed
6:	$$HE = \begin{bmatrix} HE_{11} & HE_{12} \\ HE_{21} & HE_{22} \end{bmatrix}$$
7:	where
8:	$HE_{11}(x, y) = I_x^2(x, y)$
9:	$HE_{22}(x, y) = I_y^2(x, y)$
10:	$HE_{12}(x, y) = I_x(x, y) \cdot I_y(x, y)$
11:	in addition $HE_{12} = HE_{21}$
12:	A Gaussian filter smooths the values of the components
13:	$$E = \begin{bmatrix} HE_{11} * H_\sigma & HE_{12} * H_\sigma \\ HE_{21} * H_\sigma & HE_{22} * H_\sigma \end{bmatrix} = \begin{bmatrix} A & C \\ C & B \end{bmatrix}$$
14:	The corner response function value is computed
15:	$V(x, y) = (A \cdot B - C^2) - \alpha(A + B)^2$
16:	Step 2: Convert the value $V(x, y)$ to binary form
17:	Application of threshold t_h converts the matrix into binary form
18:	$U(x, y) = V(x, y) > t_h$
19:	Step 3: Identification of salient corner features. A binary mask matrix $S(x, y)$ is generated, marking corner positions with 1 (significant) and non-corner regions with 0

(continued)

(continued)

20:	A neighborhood region is defined $R(x, y)$
21:	Initializes $S(x, y)$ with all values set to zero
22:	**for** all image coordinates (x, y) **do**
23:	**if** $(U(x, y) == 1)$ **then**
24:	**if** $(V(x, y) \geq$ to each of the values of $R(x, y))$ **then**
25:	$S(x, y) = 1$
26:	end **for**

As outlined in Algorithm 7.1, the initial step involves filtering the original image to reduce potential noise. A basic 3×3 low-pass averaging filter can effectively serve this purpose. After applying the filter, the next stage consists of calculating the image gradients in both horizontal and vertical directions. To achieve this, several filtering techniques can be employed, starting with basic operators like:

$$H_x = \begin{bmatrix} -0.5 & \underline{0} & 0.5 \end{bmatrix} \quad \text{y} \quad H_y = \begin{bmatrix} -0.5 \\ \underline{0} \\ 0.5 \end{bmatrix} \tag{7.14}$$

or Sobel

$$H_x = \begin{bmatrix} -1 & 0 & 1 \\ -2 & 0 & 2 \\ -1 & 0 & 1 \end{bmatrix} \quad \text{y} \quad H_y = \begin{bmatrix} -1 & -2 & -1 \\ 0 & 0 & 0 \\ 1 & 2 & 1 \end{bmatrix}. \tag{7.15}$$

Additional gradient computation methods and filter options are discussed in Chap. 6. Following gradient extraction, the structure matrix components are computed per Algorithm 7.1's lines 6–11.

Using the HE components, E is computed through Gaussian filtering (lines 12–13). As detailed in Sect. 5.4.1, this operation employs a coefficient matrix representation (see Fig. 7.4).

Using these computed values, we determine the corner response magnitude for every pixel as specified in algorithm line 15.

With a suitable selection of the threshold value t_h, the binary matrix $U(x, y)$ is created following the process described in lines 17 and 18. At this stage of the algorithm, the matrix $U(x, y)$ elements marked with ones $V(x, y)$ represent candidate corner points, as their corresponding values exceed the established threshold t_h. However, due to the algorithm's sensitivity regulated by the α parameter and the presence of image noise, it is common to find multiple nearby points marked as corners. Among these, the one with the highest value $V(x, y)$ is considered the actual corner. To refine this selection, step 3 of the algorithm (lines 19 to 26) is applied. In

Fig. 7.4 Gaussian
smoothing filter application

0	1	2	1	0
1	3	5	3	1
2	5	9	5	2
1	3	5	3	1
0	1	2	1	0

this phase, the goal is to identify "true corner points" by finding the local maxima $V(x, y)$ within a defined neighborhood region $R(x, y)$ centered on each candidate. Figure 7.5 presents a series of images showing the intermediate outcomes of applying Algorithm 7.1 to detect corners in a synthetic image [5].

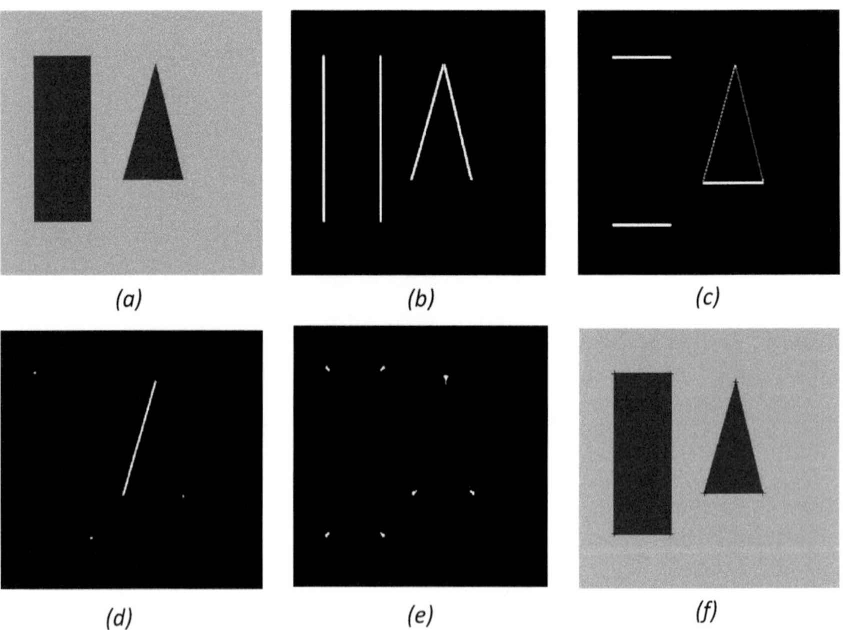

Fig. 7.5 Demonstrates the Harris algorithm applied to a synthetic image, showing intermediate outputs: **a** source image, **b–e** processing stages, and **f** final result with ground truth corner markers

7.3 Python-Based Corner Point Detection

Python does not have any function that allows us to extract the corners of an image directly, so in this case, we will implement the program that will allow us to perform this process. Following Algorithm 7.1's specifications, the Harris operator is programmed in Python as follows:

CORNER DETECTION USING HARRIS ALGORITHM

#CODE 7.1

```python
#The libraries necessary for image processing are loaded.
import cv2 as cv
import numpy as np
import matplotlib.pyplot as plt
from skimage.feature import corner_harris
from google.colab.patches import cv2_imshow

#Load image for processing
img = cv.imread('7.png')
img = cv.resize(img,dsize=None,fx=0.3,fy=0.3)
cv2_imshow(img)
```

```python
#The image is transformed from color to grayscale
imgGray = cv.cvtColor(img,cv.COLOR_BGR2GRAY)
cv2_imshow(imgGray)
```

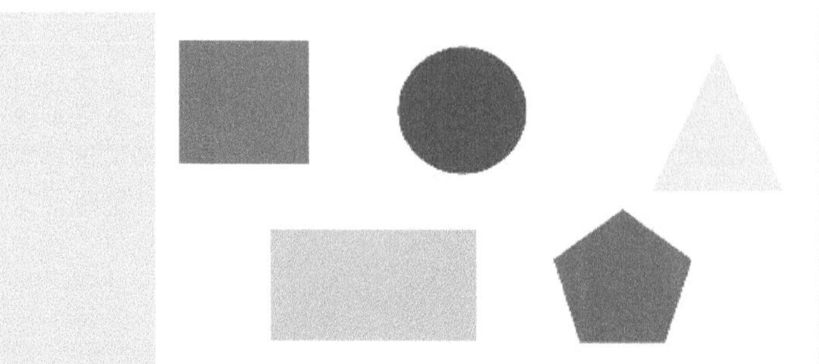

```
#A 3x3 filter box is applied
imgGrayBlur = cv.blur(np.double(imgGray),[3,3])
cv2_imshow(np.uint8(imgGrayBlur))
```

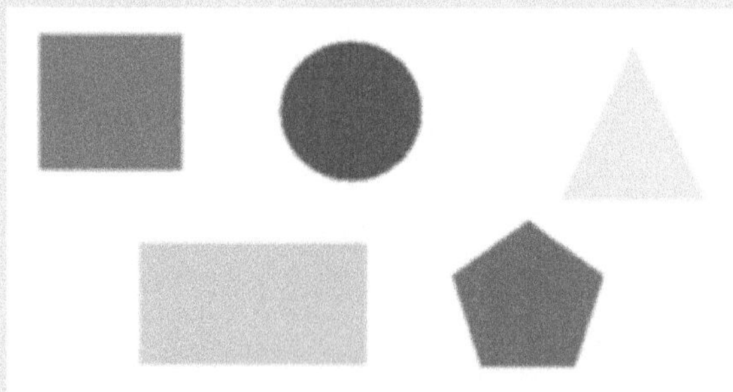

```
#Calculate gradients in x and y using sobel filter

Sx = cv.Sobel(imgGrayBlur,ddepth=cv.CV_64F,dx=1,dy=0,ksize=3)
Sy = cv.Sobel(imgGrayBlur,ddepth=cv.CV_64F,dx=0,dy=1,ksize=3)

#Calculate the elements of the HE structures matrix

Sx2 = Sx**2
Sy2 = Sy**2
Sxy = Sx*Sy
```

```python
Sx2 = Sx**2
Sy2 = Sy**2
Sxy = Sx*Sy



# Apply Gaussian filter 5x5

A = cv.GaussianBlur(Sx2,ksize=(5,5),sigmaX=0)
B = cv.GaussianBlur(Sy2,ksize=(5,5),sigmaX=0)
C = cv.GaussianBlur(Sxy,ksize=(5,5),sigmaX=0)



# Calculate the corner value function V(x,y)
# The function is equal to the determinant of the structure matrix minus its squared trace: V = det(HE)
- a*( trace (HE))**2



# A value between 0.04 and 0.25. The larger the value, the lower the sensitivity to corner detection.
a = 0.1
V = ((A*B) - (C**2)) - (a*(A+B)**2)



# Determine potential corner points with a threshold


#Threshold whose value varies between 900 and 100,000
th = 10000
U = (V > th).astype(int)
cv2_imshow(np.uint8(U*255))
```

Define the function for the maximum filter 3x3

```python
def maxFilter(V,U):
  vector = []
  [rows, cols] = V.shape
  S = np.zeros([rows,cols])
  for i in range(rows-2):
    i += 1
    for j in range(cols-2):
     j += 1
     if(U[i,j]):
      vector.append(V[i-1, j-1])
      vector.append(V[i - 1, j])
      vector.append(V[i-1, j + 1])
      vector.append(V[i, j - 1])
      vector.append(V[i, j])
      vector.append(V[i, j + 1])
      vector.append(V[i + 1, j - 1])
      vector.append(V[i + 1, j])
      vector.append(V[i + 1, j + 1])
      max = np.max(vector)
      if(V[i,j] == max):
       S[i, j] = 1
      vector = []
  return S
```

```
# Select the pixels whose V(x,y) value is the most
within a neighborhood of 8. Apply a kind of maximum 3x3 filter

S = maxFilter(V,U)
cv2_imshow(np.uint8(S*255))
```

```
img2 = img.copy()
coor = np.argwhere(S)
color = (0, 0, 255)
for i in range(coor.shape[0]):
  cv.circle(img2, np.flip(coor[i,:]),radius=3,color=color,thickness=-1)

# Flip coordinates are required because argwhere gives them upside-down

cv2_imshow(img2)
```

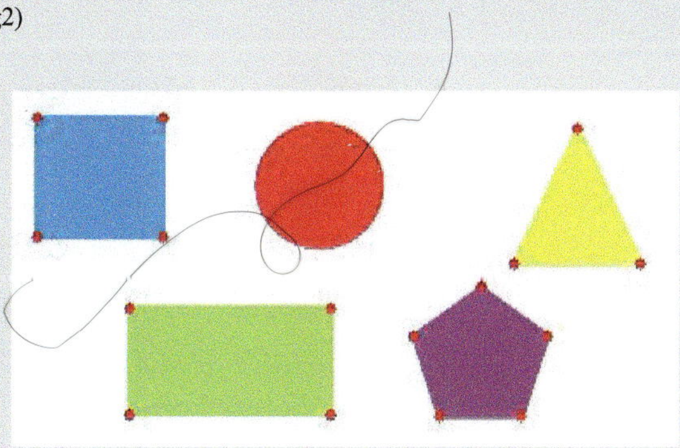

Corner Detection with OpenCV Function cv.cornerHarris()

#CODE 7.2

```python
harris = cv.cornerHarris(imgGray,blockSize=3,ksize=3,k=0.1)
harris = cv.dilate(harris,None)
imgHarris = img.copy()
imgHarris[harris > 0.0001*harris.max()] = [0,0,255]
cv2_imshow(imgHarris)
```

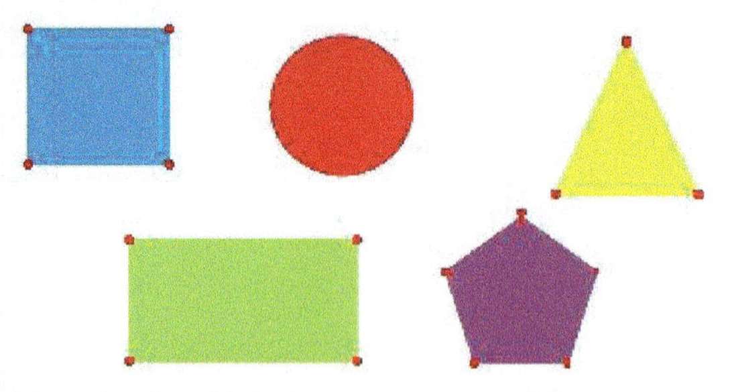

```python
# Another way to mark corners without using cv.dilate

harris = cv.cornerHarris(imgGray,blockSize=3,ksize=3,k=0.1)
harris2 = (harris > 0.001*harris.max()).astype(int)
coor = np.argwhere(harris2)
color = (0, 0, 255)
imgHarris = img.copy()
for i in range(coor.shape[0]):
  cv.circle(imgHarris, np.flip(coor[i,:]),radius=3,color=color,thickness=-1)

# Flip coordinates are required because argwhere gives them upside down
```

```
cv2_imshow(imgHarris)
```

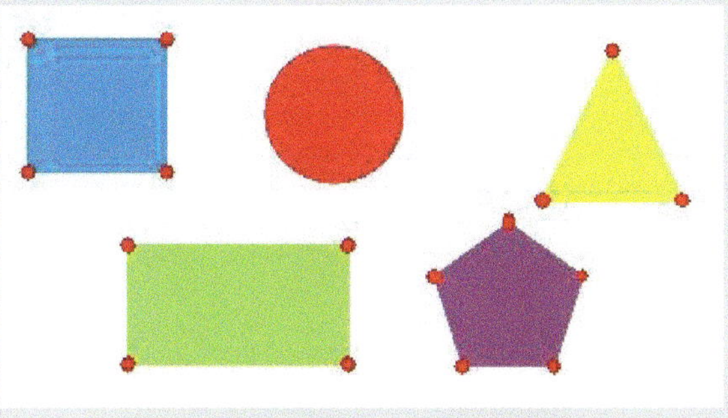

CORNER DETECTION WITH SCIKIT-IMAGE CORNER_HARRIS()

#CODE 7.3

```
harris2 = corner_harris(imgGray, k = 0.1)
harris2 = cv.dilate(harris2,None)
imgHarris2 = img.copy()
imgHarris2[harris2 > 0.0001*harris2.max()]=[0,0,255]
cv2_imshow(imgHarris2)
```

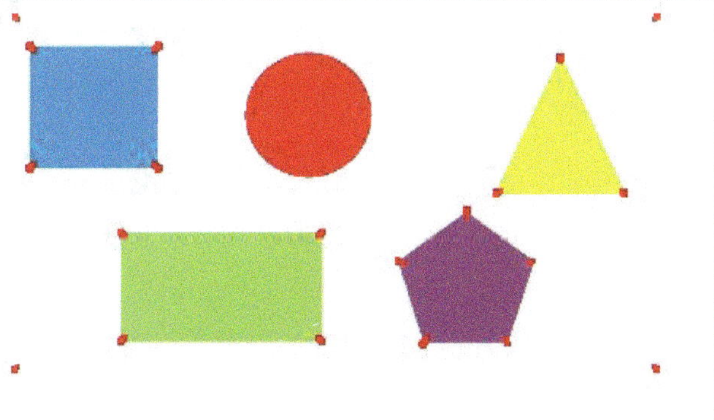

References

1. Chen, J., Zou, L. H., Zhang, J., & Dou, L. H. (2009). The comparison and application of corner detection algorithms. *Journal of multimedia, 4*(6)
2. Lavín-Delgado, J. E., Gómez-Aguilar, J. F., Urueta-Hinojosa, D. E., Zamudio-Beltrán, Z., & Alanís-Navarro, J. A. (2024). An efficient technique for object recognition using fractional Harris-Stephens corner detection algorithm. *Multimedia Tools and Applications, 83*(8), 23173–23199.
3. Mehrotra, R., Nichani, S., & Ranganathan, N. (1990). Corner detection. *Pattern recognition, 23*(11), 1223–1233.
4. Ye, Z., Pei, Y., & Shi, J. (2009). An improved algorithm for Harris corner detection. in *2009 2nd international congress on image and signal processing*, pp. 1–4. IEEE
5. Cuevas, E., Díaz-Cortes, M. A., & Mezura-Montes, E. (2019). Corner detection of intensity images with cellular neural networks (CNN) and evolutionary techniques. *Neurocomputing, 347*, 82–93.

Chapter 8
Detection of Lines and Curves

This chapter presents a comprehensive study of methods used to identify geometric structures in digital images, focusing particularly on Hough Transform and its applications in detecting lines, circles, and other parametric shapes. It begins by explaining how gradient-based edge maps, obtained from previous filtering processes, can reveal structural elements that define objects within an image. The chapter emphasizes the role of parametric models for efficiently representing geometric figures such as lines, circles, and ellipses using a small set of mathematical parameters, enabling robust identification even in noisy or fragmented data. It introduces the Hough Transform as a global detection technique that maps image points into a parameter space, where patterns of accumulation indicate the presence of specific shapes. Detailed derivations illustrate the mathematical formulations for line detection in both Cartesian (k, d) and polar (r, θ) representations, highlighting how the latter avoids singularities for vertical lines. Algorithmic implementations of line detection are provided step by step, accompanied by visual examples and Python code, including OpenCV-based versions for practical execution. The chapter extends the same framework to circle detection, formulating the problem in three-dimensional parameter space (u, v, ρ) and analyzing its computational complexity.

Chapter 6 shows how, with the help of appropriate filters, the magnitude of the gradient and its orientation in a given image pixel can be found. From applying these filters, it can be decided (perhaps by applying a certain threshold) whether or not a pixel belongs to the edge of an object. Considering the binary map obtained, lines, circles, or other curves will appear that will geometrically describe the object type contained in the image. The subject of this chapter is to find possible shapes contained in a produced binary image that could be attributed to previously specified object types.

The detection of geometric figures or parametric shapes plays a crucial role in image processing, as it enables the identification and interpretation of meaningful structures within visual data. Many real-world objects or features—such as roads, pipes, coins, signs, or anatomical structures—can be approximated or precisely

© The Author(s), under exclusive license to Springer Nature Switzerland AG 2026

E. Cuevas et al., *Image Processing with Python*, Signals and Communication Technology, https://doi.org/10.1007/978-3-032-13285-7_8

described using basic geometric forms like lines, circles, ellipses, and rectangles. By using parametric models, these shapes can be defined with a small set of mathematical parameters, allowing for efficient representation, detection, and analysis.

The importance of this approach lies in its ability to simplify complex visual scenes by reducing raw pixel data into higher-level descriptions that are more manageable and interpretable. For instance, identifying straight lines using the Hough Transform can help reconstruct the edges of buildings or detect lanes in autonomous driving systems. Similarly, detecting circles or ellipses enables applications in industrial inspection, medical imaging (e.g., locating the optic disk or tumors), and object tracking.

Parametric detection is also highly valuable for its robustness in dealing with noisy or incomplete data. Even when shapes are partially occluded or degraded, fitting a parametric model allows the system to infer the presence and properties of the underlying structure. Moreover, once geometric primitives are identified, they serve as the foundation for more advanced tasks such as object recognition, scene reconstruction, shape analysis, and feature matching.

In summary, geometric and parametric detection transforms low-level image data into structured information, facilitating accurate, efficient, and meaningful interpretation of visual content across a wide range of image processing and computer vision applications.

8.1 Detection of Structural Elements in Images

In image processing, structures refer to meaningful and coherent patterns or shapes within an image that represent distinct elements or regions, such as edges, lines, curves, corners, or complete objects [1]. These structures are typically formed by groups of pixels that exhibit consistent spatial or intensity relationships, making them visually or geometrically significant. For example, the contours of an object, the boundary of a region, or the repetitive patterns in textures can all be considered structures. Identifying and analyzing these structures are fundamental for tasks such as object recognition, segmentation, and feature extraction, as they provide crucial information about the content and organization of the image. Detecting structures often involves techniques such as edge detection, region growing, and shape modeling, which help isolate and interpret the underlying forms that define the image's composition.

An intuitive method for identifying structures in an image involves starting from a known point on an edge and progressively adding neighboring pixels that also belong to the edge, eventually tracing out the complete structure. This step-by-step accumulation can be applied to images processed through gradient thresholding or basic segmentation techniques, where edges are typically highlighted. However, this approach faces significant limitations in practice. It does not account for discontinuities or branching along the edges, which often occur due to image noise, artifacts, or the inherent limitations of the edge detection or segmentation algorithms.

These algorithms typically operate without considering higher-level shape information, making them unable to resolve ambiguities where edges are broken, faint, or complex. As a result, the process can lead to incomplete or fragmented representations of the underlying structures, reducing the accuracy and reliability of the edge-based reconstruction.

An alternative strategy for detecting structures in an image involves performing a global search to identify shapes that approximate or correspond to a predefined geometric form. Unlike local edge-following techniques, this approach aims to recognize entire structures based on their overall shape, regardless of fragmentation or noise. As illustrated in Fig. 8.1, even when numerous distracting elements, such as random dots, are added to an image, the human visual system can still effortlessly perceive and distinguish the underlying structures. However, the biological processes that enable humans and animals to recognize and relate visual patterns remain largely unknown. From a computational perspective, one effective method for addressing this challenge is the Hough Transform. This technique systematically searches for instances of specific shapes—such as lines, circles, or ellipses—by transforming image data into a parameter space where potential shapes become more easily detectable. The Hough Transform provides a robust framework for structure recognition, especially in noisy or incomplete data, and will be explored in detail throughout this chapter.

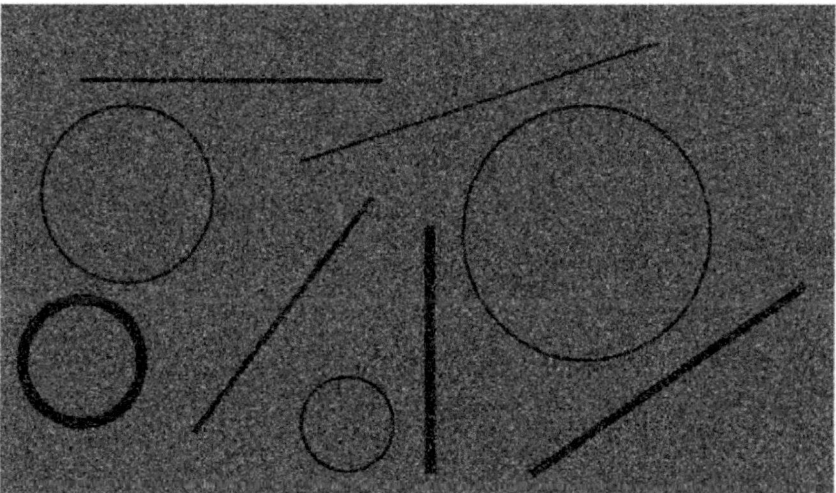

Fig. 8.1 Biological mechanisms in humans enable accurate recognition of parametric shapes like lines and circles, even when the image contains numerous additional points acting as noise, and this identification occurs with remarkable certainty

8.2 Detecting Parametric Shapes with the Hough Transform

In image processing, parametric geometric objects are shapes that can be precisely described using a set of mathematical parameters [2]. These objects include basic geometric forms such as lines, circles, ellipses, and curves, which are defined by specific equations involving a limited number of variables. For example, a straight line can be represented by its slope and intercept or by its distance and angle in polar coordinates, while a circle is defined by the coordinates of its center and its radius. These parametric representations are essential for detecting and analyzing structured elements within an image, especially in tasks like shape recognition, feature extraction, and object detection. Algorithms such as the Hough Transform rely on these models to identify instances of geometric objects by mapping image features—such as edge points—into a parameter space where the presence of specific shapes can be inferred from patterns of accumulation. This approach allows for robust and efficient detection of geometric primitives, even in noisy or partially occluded images.

The Hough Transform, originally developed by Paul Hough and patented in the USA, is widely recognized in the computer vision community as a powerful technique for detecting geometric shapes in images. This method is particularly effective for identifying parametric shapes—such as lines, circles, and ellipses—which can be mathematically described using a small set of parameters. The core idea of the Hough Transform is to analyze a distribution of points within an image and determine whether these points align in a way that forms one of the target shapes. Because lines, circles, and ellipses appear frequently in real-world images, the ability to automatically detect them is highly valuable for a range of applications, including object recognition, image analysis, and industrial inspection. As illustrated in Fig. 8.2, the transform plays a crucial role in bridging scattered image data with structured geometric information.

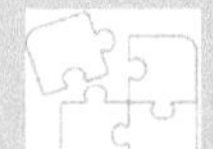

The Hough Transform is a powerful technique widely used in image processing and computer vision for detecting geometric shapes in images, particularly under challenging conditions such as noise, occlusion, or fragmentation. Its primary application is the detection of parametric shapes, including lines, circles, ellipses, and more complex forms through its generalized versions. One of the most common uses is in edge-based shape recognition, where the Hough Transform can identify lines or circular patterns by transforming edge points into a parameter space and detecting clusters or local maxima.

In industrial inspection, the Hough Transform is employed to detect structural defects, misalignments, or regular patterns—such as circular holes, bolts, or straight edges—on manufactured products. In medical imaging, it helps

Fig. 8.2 Frequent occurrence of lines, circles, and ellipses in images

identify anatomical structures like blood vessels or circular features such as optic disks in retinal images. In robotics and autonomous vehicles, the transform is used for lane detection, where lines on roads are extracted from camera feeds to assist in navigation.

Another application is in document analysis, where the Hough Transform aids in detecting lines of text, page margins, or skew correction by identifying consistent linear structures. In astronomy, it is applied to detect circular celestial objects or star trails in images. The Hough Transform also finds use in augmented reality and pattern recognition, supporting feature matching and scene understanding by locating predefined geometric patterns.

Its ability to robustly detect shapes, even with partial data or heavy noise, makes the Hough Transform an essential tool in a wide range of domains, from automated systems and scientific analysis to visual interpretation and real-time processing tasks.

We begin by examining the application of the Hough Transform for detecting lines in binary images, which are typically generated through gradient thresholding—a common preprocessing step in many computer vision and image processing systems. This specific use of the Hough Transform is one of its most frequent and practical implementations, as it allows for the robust identification of linear structures within edge-detected images. In a two-dimensional space, a straight line can be mathematically represented using two parameters, as defined in Eq. 8.1. These parameters form the basis for mapping image points into a parameter space, where patterns corresponding to lines become more apparent and easier to detect, even in the presence of noise or incomplete edges.

$$y = kx + d. \tag{8.1}$$

In the equation, the parameter k represents the slope of the line, while d denotes the point at which the line intersects the y-axis, also known as the y-intercept (as illustrated in Fig. 8.3). These two parameters define the orientation and position of the line within the image plane. For any line that passes through two distinct points, $p_1 = (x_1, y_1)$ and $p_2 = (x_2, y_2)$, it must satisfy the linear relationship outlined in Eq. 8.2. This condition ensures that both points lie precisely on the same straight line, forming the foundation for determining whether a set of image points can be associated with a common line during the Hough Transform process.

$$y = kx_1 + dy \; yy = kx_2 + d \tag{8.2}$$

In the line equation, the parameters k and d belong to the set of real numbers R. The goal is to estimate these parameters for a line that passes through multiple points located along the edges of an object in an image. This leads to a key question: how can

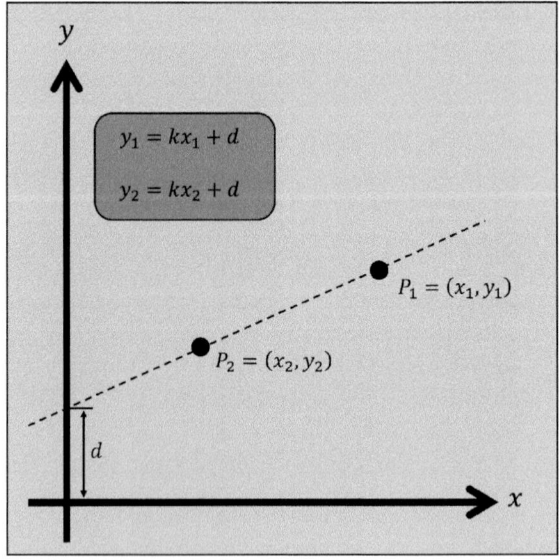

Fig. 8.3 Since the two points p_1 and p_2 lie on the same line, determining the equation of that line requires estimating the values of k (slope) and d (y-intercept)

we determine which lines actually correspond to real edge structures, and how many edge points a given line contains? A straightforward but computationally intensive approach would be to generate every possible line in the image, count how many edge points lie on each one, and then retain only those lines that intersect a significant number of points. While this brute-force method is conceptually simple, it is highly impractical due to the vast number of potential lines that could be drawn, especially when considering subpixel variations and real-valued parameters. The sheer volume of computations required would result in extreme inefficiency, highlighting the need for a more optimized solution—such as the Hough Transform—to perform this task effectively.

Using a parametric model of the line equation is essential for detecting lines in images because it provides a precise and compact mathematical representation that facilitates systematic analysis and robust detection. By expressing a line through parameters—such as slope and intercept in Cartesian form, or distance and angle in polar form—edge points in an image can be transformed into a parameter space where patterns can be more easily identified. This is particularly useful in noisy or fragmented images where lines may not appear as continuous sequences of pixels. Parametric models enable algorithms like the Hough Transform to accumulate evidence from multiple edge points and identify consistent alignments that correspond to actual lines. Without such a model, detecting lines would rely solely on pixel connectivity or heuristics, which are less accurate and more sensitive to noise. Therefore, the parametric approach not only improves computational efficiency but also increases robustness and accuracy in identifying linear structures within complex image data.

8.2.1 Representation in Parameter Space

The Hough Transform approaches the problem of line detection in a more efficient and elegant way by generating all possible lines that could pass through a given pixel $P_0 = (x_0, y_0)$, which corresponds to an edge point in the image. Instead of analyzing every possible line across the image space [3], the transform focuses only on edge pixels and calculates the set of all lines that could pass through each one. Each of these lines, denoted as L_p, follows a mathematical model defined by Eq. 8.3. This equation describes the relationship between the parameters of the line—typically slope and intercept or an equivalent form—and the coordinates of the edge point. By mapping each edge point to its corresponding set of potential lines in parameter space, the Hough Transform efficiently accumulates evidence for the presence of lines based on the intersections of these mappings, significantly reducing computational complexity while maintaining robustness to noise and fragmentation.

$$L_p : y_0 = \mathrm{k}x_0 + d. \tag{8.3}$$

In this context, the values of k (slope) and d (y-intercept) are systematically varied to generate all possible lines that pass through the fixed point (x_0, y_0). This process results in a set of solutions to Eq. 8.3 that represents an infinite number of lines intersecting at that single point. Each unique combination of k and ddd defines a distinct line sharing the common coordinates x_0 and y_0. As illustrated in Fig. 8.4, this collection of lines forms a family of solutions in parameter space, all anchored to the same image point. For any fixed value of k, Eq. 8.3 can be rearranged to compute the corresponding value of d, describing a curve in the (k, d) parameter space. This functional relationship is formally expressed in Eq. 8.4 and forms the basis of the Hough Transform's strategy for accumulating evidence of line presence in the parameter domain.

$$d = y_0 - \mathrm{k}x_0. \tag{8.4}$$

Equation 8.4 represents a linear function in which k and d are treated as variables, while x_0 and y_0 are fixed constants corresponding to the coordinates of a specific point $p_0 = (x_0, y_0)$ in the image. This equation defines a straight line in the (k, d) parameter space, and the set of all solutions $\{(k, d)\}$ describes every possible pair of slope and intercept values that define a line L_p passing through the point p_0. In other words, each point on this line in parameter space corresponds to a different geometric

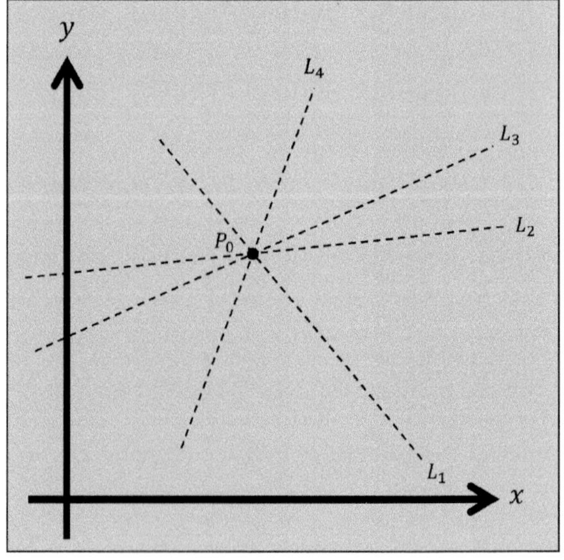

Fig. 8.4 Collection of lines intersecting at a common point p_0: every possible line L_p that passes through p_0 can be described by the equation $y_0 = \mathrm{k}x_0 + d$, where k and d are the variable parameters that distinguish one line from another

line in image space that intersects p_0, making this representation fundamental to the Hough Transform's process of identifying line patterns based on point-to-parameter mappings.

For any given pixel in the image, denoted as $p_i = (x_i, y_i)$, Eq. 8.4 establishes a relationship between the slope k and intercept d for lines passing through that point. Based on this relationship, each pixel corresponds to a continuous set of lines whose parameters satisfy this condition. This set of lines is formally expressed by Eq. 8.5, which defines all possible combinations of k and d that result in lines intersecting the specific pixel p_i. In the context of the Hough Transform, this means that every edge pixel contributes a curve in the parameter space, and the intersection of such curves from multiple pixels indicates the presence of a common line in the image.

$$R_i : d = y_i - kx_i \tag{8.5}$$

The parameters k and d serve as the variable components that define the parameter space, also known as the Hough space. This space is used to represent the different possible lines in an image based on their slope and intercept. In contrast, x_i and y_i represent fixed values corresponding to the coordinates of a pixel in the original image, forming what is known as the image space or image parameter space. While the image space describes the spatial location of edge points, the Hough space captures all potential geometric configurations (lines) that could pass through those points. Table 8.1 provides a summary of how these two spaces are conceptually and mathematically related, highlighting the transformation between the pixel domain and the parameter domain used in the Hough Transform.

In the Hough Transform framework, each point p_1 in the image parameter space (i.e., the image's pixel domain) corresponds to a line in the Hough parameter space (defined by the variables k and d). This transformation is based on the idea that any point in the image could lie on an infinite number of lines, and each of those lines is represented as a point in Hough space. The key insight is that when multiple such lines in Hough space intersect at a single point, it indicates that the corresponding points in the image space all lie on the same straight line. In other words, the point of intersection in Hough space represents the parameters (k, d) of a line that passes through all contributing edge points in the image. As illustrated in Fig. 8.5, the intersection of lines R_1 and R_2 at the point $q = (k_{12}, d_{12})$ in Hough space corresponds to two points, p_1 and p_2, in the image space. Therefore, the detected line in the image will have a slope k_{12} and a y-intercept d_{12}. The more lines that intersect at a given point in the Hough space, the greater the number of image points that align along the

Table 8.1 Existing relationship of points and lines between the different parameter spaces

Spatial domain of the image (x, y)	Hough space representation (k, d)
Point: $p_i = (x_i, y_i)$	Line: $R : d = y_i - kx_i$
Line: $L : y_i = k_j x + d_j$	Point: $q_j = (k_j, d_j)$

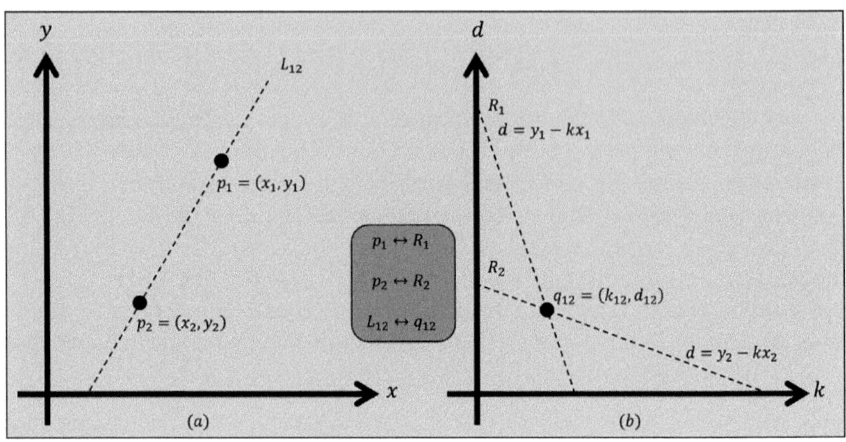

Fig. 8.5 Correspondence between image space and Hough space: **a** Image parameter space and **b** Hough parameter space

same line—indicating a stronger, more reliable detection of that line in the image. These elements can be stated as:

If NR represents the number of lines that intersect in $(\underline{k}, \underline{d})$ of the Hough parameter space, then there will be NR points that lie on the line defined by $y = \underline{k}x + \underline{d}$ in the image space

---

Figure 8.5 illustrates the fundamental principle behind the Hough Transform: the dual mapping between the image parameter space and the Hough parameter space. In this relationship, each point in the image space (typically representing an edge pixel) corresponds to a curve—often a line—in the Hough space, where all possible geometric shapes that could pass through that point are represented. Conversely, each point in the Hough parameter space defines a specific geometric shape (such as a line) in the image. For example, the points p_1 and p_2 in the image parameter space map to the lines R_1 and R_2 in the Hough space, representing all the lines that could pass through p_1 and p_2, respectively. Where these lines intersect—at point q_{12} in the Hough space—indicates a set of parameters that define a real line L_{12} passing through both p_1 and p_2 in the image. This dual correspondence between spaces enables the Hough Transform to robustly detect geometric shapes by identifying clusters or intersections in parameter space that correspond to actual structures in the image.

8.2.2   Voting Matrix in Hough Space

The process of detecting lines in an image using the Hough Transform relies on identifying points in the Hough parameter space where multiple lines intersect, as

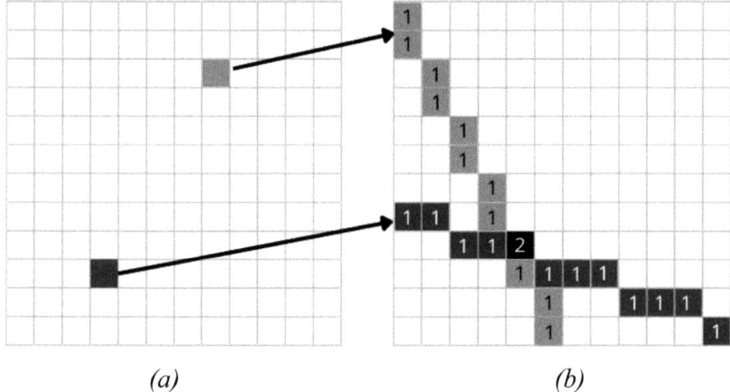

Fig. 8.6 Core concept of the Hough Transform: **a** Image parameter space and **b** accumulator in the Hough parameter space

these intersections indicate a strong presence of a line in the original image. To compute the Hough Transform, the continuous ranges of the parameters k (slope) and d (intercept) must first be discretized into a grid of finite steps, allowing for practical implementation. Each possible line in the image—derived from an edge pixel—is mapped into the Hough space as a line, and the algorithm keeps track of how many such lines intersect at each coordinate in this space. This is done using an accumulator matrix, where each cell represents a discrete pair (k, d), and is incremented every time a corresponding line passes through it. The value stored in each cell, N_p, thus indicates how many pixels (or points) in the image contribute to a particular line defined by that (k, d) pair. A high N_p value suggests that many image pixels align with that line, signaling its probable existence. This accumulation process is illustrated in Fig. 8.6, which builds upon the example shown in Fig. 8.5, making the Hough Transform a powerful tool for robust line detection even in noisy or fragmented images.

8.2.3 Modifying the Parametric Representation

Up to this point, we have explored the fundamental concept behind the Hough Transform and its application in detecting lines by mapping image points to a parametric space. However, the initial line representation given by Eq. 8.1, which expresses a line using slope k and intercept d, presents a critical limitation: it becomes problematic when dealing with vertical lines. Specifically, vertical lines correspond to an undefined or infinite slope, and setting $k = 0$ leads to computational inaccuracies or errors. To overcome this issue, a more robust and numerically stable alternative is adopted, as defined in Eq. 8.6. This revised formulation avoids the division by zero associated with vertical lines and allows for a consistent representation of all line orientations, making it better suited for practical implementation of the Hough Transform in digital image processing.

$$x \cdot \cos\cos(\theta) + y \cdot \text{sen}(\theta) = r. \tag{8.6}$$

Equation 8.6 offers a significant advantage in that it does not exhibit any singularities, making it well-suited for computational applications involving line detection. Unlike previous formulations that become unstable or undefined for certain cases—such as vertical lines—this equation remains consistent across all orientations. Additionally, it enables linear quantization of its parameters r (the distance from the origin to the line) and θ (the angle the line makes with the x-axis), allowing the parameter space to be discretized uniformly. This uniform quantization simplifies the implementation of the Hough Transform, facilitates efficient accumulation in the parameter space, and ensures that all possible lines, regardless of direction or position, are treated equally during detection.

The accumulator matrix [4] in Fig. 8.6 serves as a discretized version of the Hough parameter space, allowing the Hough Transform to be implemented computationally. For each edge point detected in the image space (as illustrated in part (a)), a corresponding curve—usually a line—is computed in the Hough parameter space. This curve represents all possible geometric configurations (e.g., slope k and intercept d) of lines that could pass through that image point. As the algorithm processes each edge point, it adds votes to the cells in the accumulator matrix where the corresponding curves pass, incrementing their values by one. This additive process builds up evidence for the presence of specific line parameters. Over time, cells with the highest vote counts, known as local maxima, emerge in the accumulator. These peaks indicate the parameter pairs (k, d) that are most frequently supported by edge points in the image and, therefore, represent the most probable lines present in the original image.

Figure 8.7 demonstrates the limitations of using Eq. 8.1 for line representation, particularly the singularity that arises when attempting to model vertical lines, where the slope becomes undefined. In Fig. 8.7, lines can be represented using different sets of parameters, but their effectiveness varies depending on the line's orientation. A common method is the slope-intercept form, which works well for most lines but fails when modeling vertical or near-vertical lines, as the slope kkk becomes infinite ($k = \infty$), leading to a singularity, as shown in option (a). To overcome this limitation, Eq. 8.6 introduces a more robust parametric model that defines lines using r and θ, where r is the perpendicular distance from the origin to the line, and θ is the angle of this normal. This approach, illustrated in option (b), eliminates singularities and supports linear quantization, making it more suitable for computational implementation. Because it provides a stable and consistent way to represent lines regardless of their orientation, it is especially well-suited for the Hough Transform, where accuracy and uniformity across all angles are essential.

To address this issue, Eq. 8.6 offers a more stable alternative for describing straight lines. When this new formulation is adopted, the structure of the Hough parameter

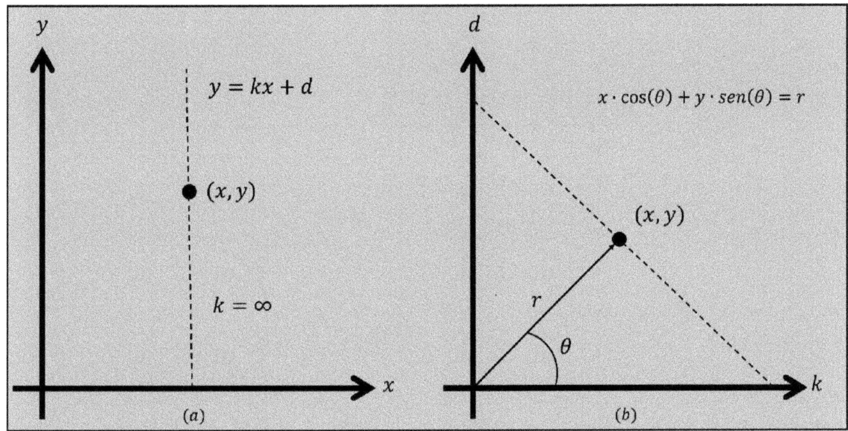

Fig. 8.7 Different parameterizations can define lines, but their effectiveness depends on orientation. The slope-intercept form Fig. 8.7a fails for vertical lines due to infinite slope ($k = \infty$),). Equation 8.6 provides a more robust alternative, using r and θ to represent a line by its normal vector. As it is shown in Fig. 8.7b, this model avoids singularities, supports linear quantization, and ensures stable detection of lines at any angle—making it ideal for use in the Hough Transform

space changes accordingly, shifting from slope-intercept parameters to a representation based on the distance r and angle θ. In this revised space, each point $p_1 = (x_i, y_i)$ in the image parameter space corresponds to a sinusoidal curve in the Hough space. The relationship between the image point and the Hough parameters is defined by Eq. 8.7, which establishes how each edge pixel contributes to the accumulation process in terms of r and θ, enabling robust detection of lines across all orientations.

$$r_{x_i, y_i} = x_i \cdot \cos\cos(\theta) + y_i \cdot \mathrm{sen}(\theta) \tag{8.7}$$

In the context of the Hough Transform using the polar coordinate representation, the parameter θ typically ranges from $0 \leq \theta < \pi$, , covering all possible line orientations in the image (as illustrated in Fig. 8.8). To handle both positive and negative coordinate values, the image center (x_c, y_c) is often chosen as the reference point for defining pixel locations. This re-centering allows pixel coordinates to span positive and negative values along both axes. As a result, the corresponding values for the parameter r, , which represents the perpendicular distance from the origin (image center) to the line, become symmetrically constrained. The maximum and minimum values of r are determined based on the image dimensions and are mathematically defined by the expression given in Eq. 8.8. This constraint ensures that the Hough parameter space fully encompasses all possible lines that can pass through the image area.

$$-r_{max} \leq r_{x,y}(\theta) \leq r_{max}, \tag{8.8}$$

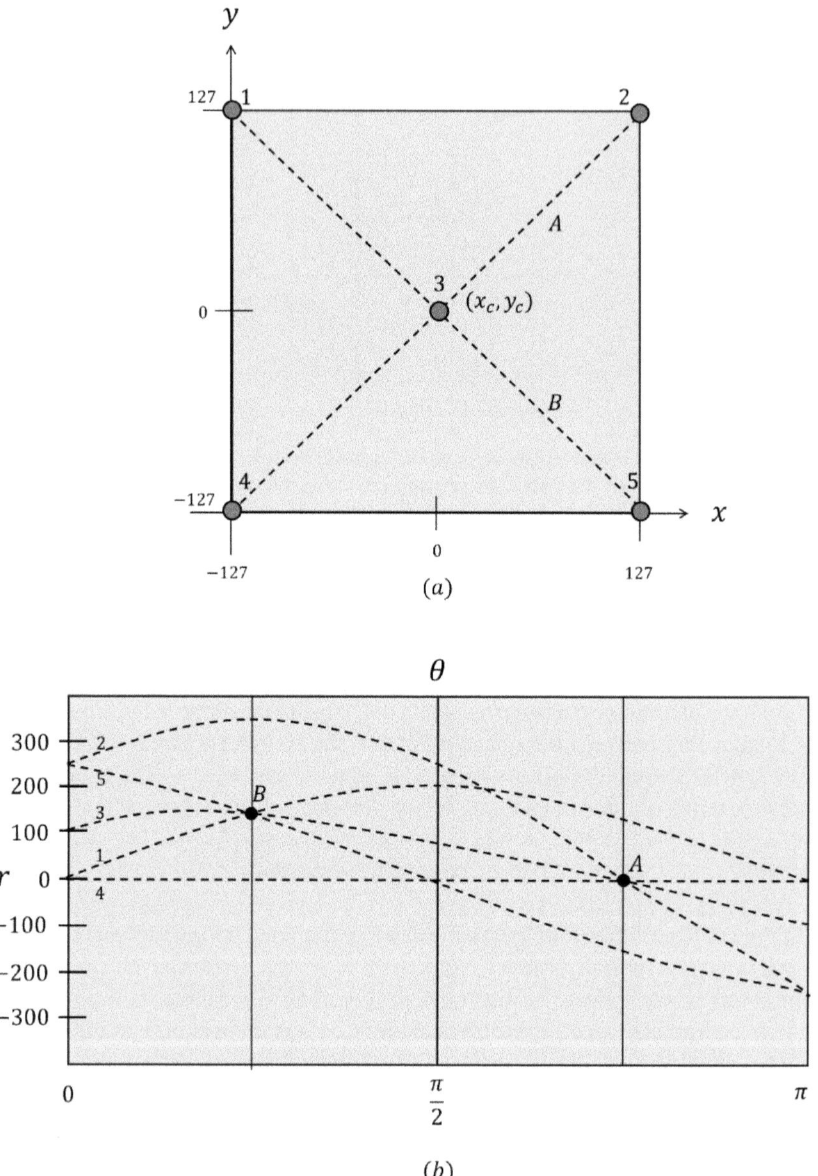

Fig. 8.8 **a** Image space and **b** Hough space based on the straight-line representation defined by Eq. 8.6

where

$$r_{\max} = \sqrt{\left(\frac{M}{2}\right)^2 + \left(\frac{N}{2}\right)^2},$$ (8.9)

where M and N denote the image's width and height, respectively.

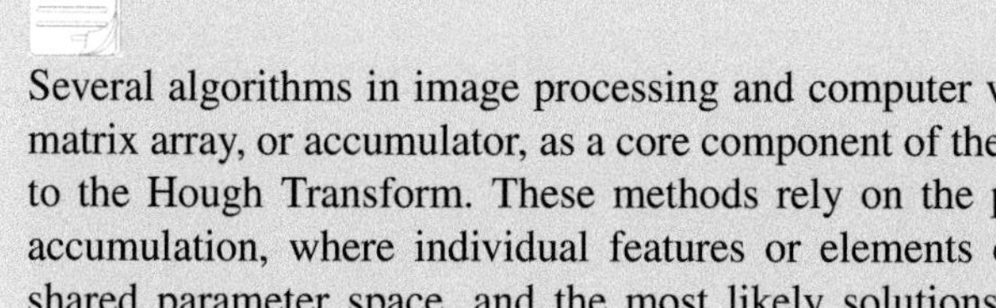

Several algorithms in image processing and computer vision utilize a voting matrix array, or accumulator, as a core component of their operation—similar to the Hough Transform. These methods rely on the principle of evidence accumulation, where individual features or elements contribute votes to a shared parameter space, and the most likely solutions emerge as peaks or local maxima. Beyond the Hough Transform for line and circle detection, the Generalized Hough Transform extends the concept to detect arbitrary shapes by learning a shape model and voting for possible transformations. In 3D vision, Random Sample Consensus (RANSAC) uses a voting-like mechanism to robustly fit geometric models (like planes or cylinders) to noisy data, identifying inliers that support a given hypothesis. Additionally, object detection techniques such as Hough Forests and Implicit Shape Models employ voting schemes to localize object parts or centers based on learned features. Even in pose estimation, methods like Hough voting for 6D object pose estimation rely on aggregating predictions to robustly determine object position and orientation. These voting-based approaches are valued for their robustness to noise, partial data, and cluttered environments, making them powerful tools for detecting patterns and structures in complex visual data.

8.3 Implementation of Line Detection via Hough Transform

Algorithm 8.1 outlines the process for detecting line parameters in an image using the Hough Transform based on the straight-line model defined by Eq. 8.6. The input to the algorithm is a binary image, which contains only two possible pixel values: ones and zeros. This binary image is typically obtained through edge detection methods, such as those discussed in Chap. 6. In this context, a pixel value of one indicates that the pixel is part of an edge, while a zero signifies background. The algorithm systematically processes all edge pixels, mapping them into the Hough parameter space using the (r, θ) model, and accumulates votes in a discretized matrix to identify the most likely line candidates. This approach enables robust and efficient detection of straight lines, even in the presence of noise or incomplete edge information.

Algorithm 8.1 Procedure for line detection using the Hough Transform

	Hough Transform Line Detector $(I(x, y))$
	$MRAcc(\theta, r) \rightarrow$ Matrix of accumulator registers
	$(x_c, y_c) \rightarrow$ Coordinates of the center of $I(x, y)$
1:	$0 \rightarrow MRAcc(\theta, r)$
2:	*for* all coordinates of the image $I(x, y)$ *do*
3:	*if* $(I(x, y)$ is an edge)*then*
4:	$(x - x_c, y - y_c) \rightarrow (u, v)$
5:	*for* $\theta_i = 0 \dots \theta$ *do*
6:	$r = u \cdot coscos(\theta_i) + v \cdot sen(\theta_i)$
7:	$MRAcc(\theta, r)$ is incremented
8:	Finally, the $MRAcc(\theta, r)$ registers whose values are maximum are found

The Hough Transform algorithm for line detection begins by initializing all cells in the accumulator matrix to zero (statement 1). This matrix serves as a register to count how often specific combinations of line parameters occur. Next, the algorithm traverses the entire binary image $I(x, y)$ (statement 3), where edge pixels—those with a value of one—are processed further. For each detected edge pixel, the algorithm uses Eq. 8.6 to compute potential line parameters by sweeping the angle θ from 0 to π (statements 5 and 6). To calculate the corresponding r values, the coordinates of each pixel are adjusted using the image center (x_c, y_c) as a reference. Each resulting (r, θ) pair identifies a specific cell in the accumulator matrix, which is incremented to reflect the presence of a potential line passing through that pixel (statement 7). After scanning the entire image, the accumulator matrix will have several peaks or local maxima (statement 8), indicating the most likely line candidates based on accumulated votes. To extract these lines, a threshold is first applied to eliminate low-value entries, retaining only significant peaks. A neighborhood-based search is then performed to identify local maxima, which corresponds to the strongest line detections. Some neighboring cells may also contain relatively high values due to noise or quantization errors introduced by the discretization of the Hough parameter space. These imperfections can cause nearby cells to accumulate partial votes for the same line, making local peak detection essential for accurately identifying the parameters (r, θ) that define real lines in the original image $I(x, y)$.

Figure 8.9 presents a sequence of images illustrating the intermediate results generated during the execution of Algorithm 8.1 for line detection using the Hough Transform. In Fig. 8.9a, the original image is shown—an artificially created example designed to demonstrate the functionality of the algorithm. Figure 8.9b displays the result of applying the Canny edge detection algorithm (discussed in Chap. 6), which extracts the prominent edges from the original image. In Fig. 8.9c, the accumulator matrix—also known as the Hough parameter space—is visualized, where bright spots indicate the cells with high counts, representing likely values of r and θ corresponding to real lines in the image $I(x, y)$. Figure 8.9d shows the result of thresholding this matrix to isolate only the most significant votes. Following that, Fig. 8.9e highlights the locations of local maxima within a defined neighborhood in the thresholded matrix, identifying the most probable parameter pairs for line detection. Finally, Fig. 8.9f displays the actual lines detected in the original image, reconstructed using the r and θ values obtained from the Hough Transform, effectively completing the line detection process.

From Fig. 8.9e, the identified maximum points reveal that points 5 and 6—located very close to each other and nearly adjacent in the Hough parameter space—correspond to the pair of parallel lines shown in Fig. 8.9a. Their proximity in the accumulator matrix indicates that these lines share a very similar orientation, specifically having the same or nearly identical value of θ, which represents the slope in the polar parameter model. This illustrates how parallel lines in the image space are mapped to nearby peaks in the Hough parameter space, since they differ mainly in their r values (distance from the origin), but maintain a common direction characterized by the same θ.

8.4 Line Detection Using the Hough Transform in Python

In the following section, we will examine two different methods for detecting lines using the Hough Transform. The first approach involves a detailed, step-by-step implementation of the standard Hough Transform algorithm, allowing us to explore each computational stage involved in identifying straight lines within a binary image. This method is particularly valuable for understanding the mathematical foundations of the transform, including how edge points are mapped into parameter space and how the accumulation process reveals prominent line features. After establishing a solid conceptual understanding, we will shift to a more practical approach by utilizing OpenCV's built-in Hough Transform function. This high-level implementation simplifies the process by automatically handling the underlying computations and directly returning the parameters of the detected lines, offering a more efficient solution for real-world applications.

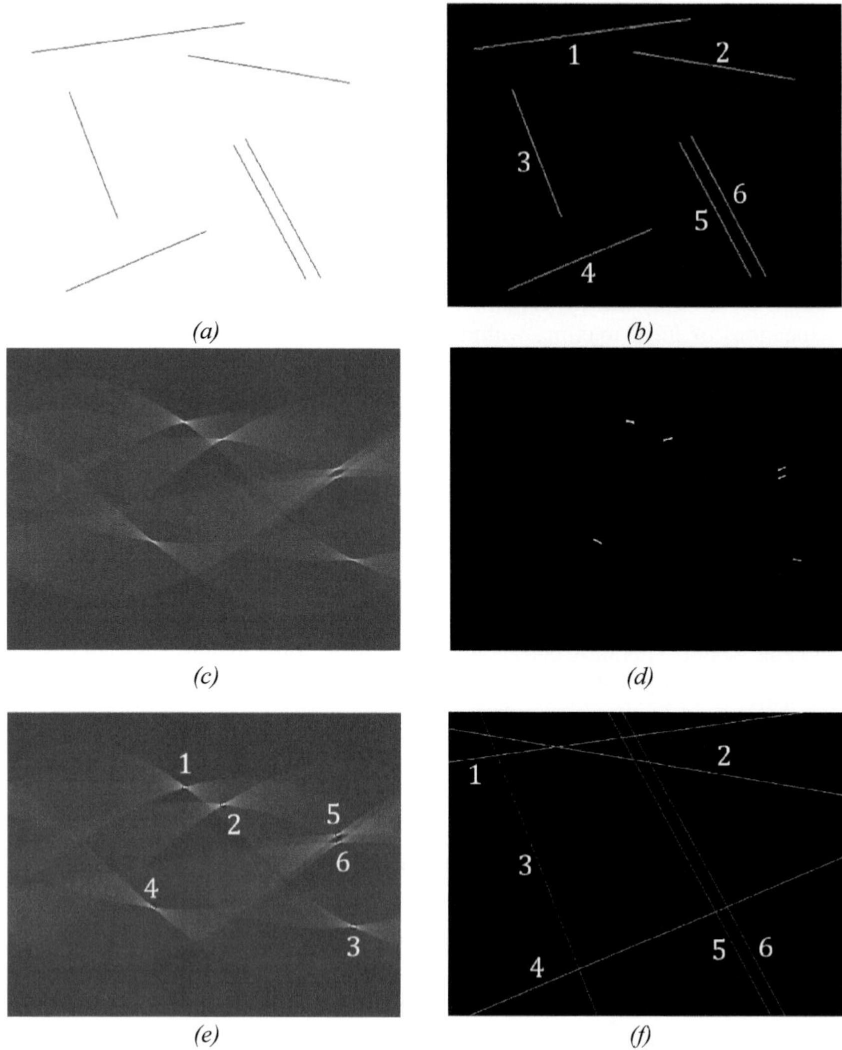

Fig. 8.9 Intermediate results from applying the Hough algorithm for line detection: **a** the original image, **b** the edge-detected version of the image, **c** the accumulator matrix representing the Hough parameter space, **d** the thresholded version of the accumulator matrix, **e** identification of peak points in the parameter space, and **f** the resulting lines detected by the Hough Transform

Hough Transform For Line Detector

#CODE 8.1

```python
# The libraries necessary for image processing are loaded.

import cv2 as cv

import numpy as np

import matplotlib.pyplot as plt

from google.colab.patches import cv2_imshow
```

```python
img = cv.imread('Image8.1.jpg')

cv_imshow(img)

imgGray = cv.cvtColor(img,cv.COLOR_BGR2GRAY)
```

```python
# Image segmentation to obtain a binary image

imgBim= cv.Canny(imgGray,threshold1=100,threshold2=200)

cv_imshow(ImgBin)
```

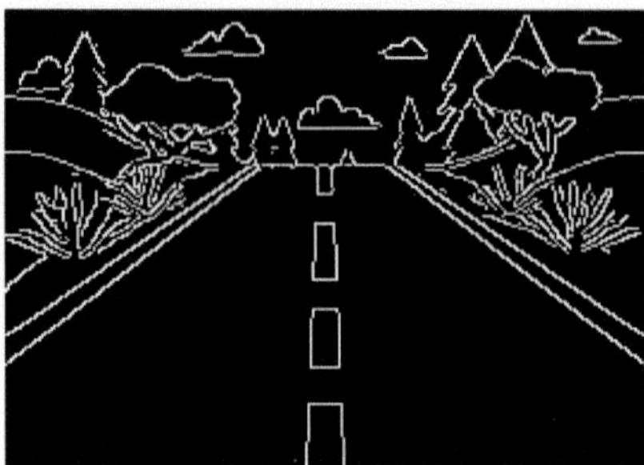

m,n = imgBin.shape

Initialize the Hough accumulator matrix with zeros

AccMatrix = np.zeros((m,n))

Compute the maximum possible value of r based on the image dimensions (see Equation 8.9)

rmax = np.sqrt((m)**2 + (n)**2)

Define linear scaling for the Hough parameters: theta and r

iA = np.pi/n   # Angular step size

ir = rmax/m   # Radial step size

Define the center of the image as the origin of the coordinate system

m2=int(m/2)

n2=int(n/2)

Iterate over the binary image, focusing on edge points (pixels with value 1)

```python
for re in range(m):

  for co in range(n):

   if imgBin[re,co]:

    for an in range(n):

      # Convert image coordinates relative to the center

      x = co - n2

      y = re - m2

      theta = an * iA


      # Compute the corresponding r value (see Equation 9.6)

      r = int((((x*np.cos(theta) + y*np.sin(theta))/ir) + m2)

        # Check if r is within the valid range

      if (r >= 0) and (r < m):

        #  Increment the corresponding accumulator cell for (r, theta)

        AccMatrix[r,an] += 1


# Normalize and display the accumulator matrix as an image

cv_imshow(np.uint8(AccMatrix/np.max(AccMatrix)*255))
```

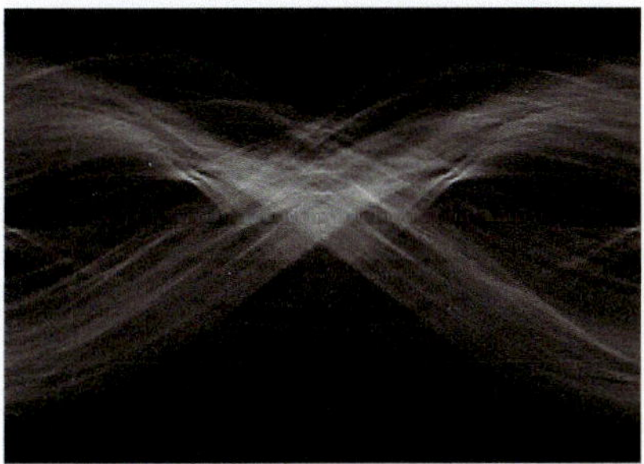

```python
# Set the threshold to identify strong line candidates in the accumulator matrix

th = 150

#th = 200          # Optional alternative threshold value


# Create a binary mask where accumulator values greater than the threshold are set to 1

Flag = (AccMatrix > th).astype(int)



# Display the binary image highlighting detected line positions

cv_imshow(np.uint8(Flag*255))
```

```python
# Define the pixel neighborhood size for local maximum search

neighborhood = 3


# Initialize the matrix to store detected local máxima

final_maxima = np.zeros((m, n))
```

```python
# Traverse the Flag matrix to find potential line points

for row in range(m):

  for col in range(n):

   if Flag[row,col] == 1:

      # Define the search neighborhood region

        row_start = max(row - neighborhood, 0)

        row_end = min(row + neighborhood, m)

        col_start = max(col - neighborhood, 0)

        col_end = min(col + neighborhood, n

        # Extract the block around the current pixel

        Block = AccMatrix[row_start:row_end, col_start:col_end]

   max_value = np.max(Block)

   # Check if the current pixel is a local máximum

   if AccMatrix[row,col] >= max_value:

   # Mark the local maximum in the final maxima matrix

    final_maxima[row, col] =1


# Display the final local maxima matrix as a binary image

cv_imshow(np.uint8(final_maxima*255))
```

```python
# Retrieve the coordinates of pixels marked as local maxima,

# which correspond to accumulator entries representing detected lines

ind = np.argwhere(final_maxima)

r_indices = indices[:, 0]

theta_indices = indices[:, 1]




# Display the detected lines

# Get the total number of detected lines (i.e., the number of local maxima)

num_lines = r_indices.size




# Initialize an empty matrix to draw the detected lines

line_image = np.zeros((m, n))




# Draw all detected lines

for i in range(num_lines):

    # Retrieve the parameters (r, theta) of each detected line

    r_idx = r_indices[i]
```

```python
theta_idx = theta_indices[i]

# Scale the values of r and theta back to their original ranges

r_value = (r_idx - m2) * ir

theta_value = theta_idx * iA

# Compute the horizontal and vertical projections of r, as r is the vector from the origin
perpendicular to the detected line

x = np.round(r_value * np.cos(theta_value))

y = np.round(r_value * np.sin(theta_value))


# Correct the offset caused by using the image center as the reference

# Trace the detected line model using the recovered parameters

# First trace in the positive direction

iterable = np.arange(start=x,stop=n2,step=1)

for c in iterable:

  r = np.round(((-(x/y)*c) + y + (x*x/y)) + m2)

  if 0 <= r < m:

    line_image[int(r),int(c+n2)] = 1


# Then trace in the negative direction

iterable = np.arange(start=x,stop=1-n2,step=-1)

for c in iterable:

  r = np.round(((-(x/y)*c) + y + (x*x/y))) + m2

  if 0 <= r < m:

    line_image [int(r),int(c+n2)] = 1
```

Display the final image with detected lines

cv_imshow(line_image * 255)

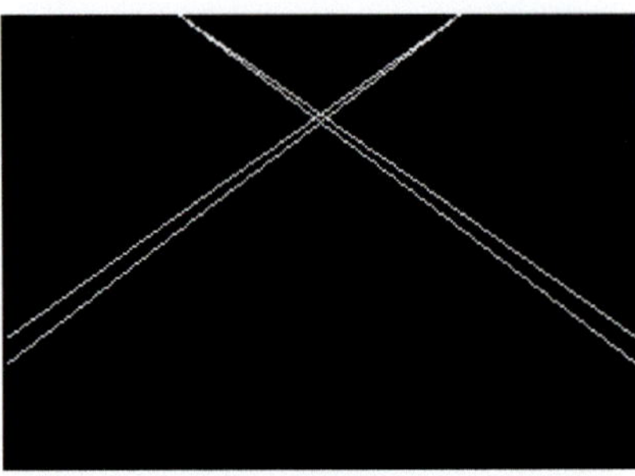

This section implements the Hough Transform using OpenCV's built-in function to detect
lines in a binary image

lines = cv.HoughLines(imgBin,rho=1, theta=np.pi/180, threshold=100)

imgHough = img.copy()

if lines is not None:

  for i in range(len(lines)):

    theta = lines[i][0][1]

    rho = lines[i][0][0]

    y0 = rho * np.sin(theta)

```
x0 = rho * np.cos(theta)


pt1 = int(x0  - 1000*np.sin(theta)), int(y0 + 1000*np.cos(theta))

pt2 = int(x0  + 1000*np.sin(theta)), int(y0 - 1000*np.cos(theta))

cv.line(imgHough, pt1=pt1, pt2=pt2, color=(0,0,255), thickness=1)

cv_imshow(imgHough)
```

8.5 Detection of Circular Features via Hough Transform

Lines are fundamental geometric primitives in two-dimensional space and can be
fully defined using just two parameters, such as slope and the intercept or distance
and angle. In contrast, a **circle**, though also a 2D shape, requires three parameters for
complete representation: the coordinates of its center, typically given as an ordered
pair (u, v), and its radius ρ. These three parameters together determine the circle's
exact position and size within the image [5]. Figure 8.10 illustrates this relationship,
showing how the center and radius define the circle geometrically and how these
parameters relate to the spatial characteristics of the primitive in the image plane. This
distinction in parameter requirements is crucial when applying detection techniques
like the Hough Transform, as it directly affects the complexity and dimensionality
of the parameter space.

As illustrated in Fig. 8.10, a point $P = (x, y)$ belongs to a circle if it satisfies the
circle equation defined by Eq. 8.10. This equation mathematically models a circle
based on its center coordinates (u, v) and radius ρ, expressing the condition that the
distance between the point P and the center of the circle is exactly equal to the radius.
If this relationship holds true, then the point lies precisely on the circumference of the
circle. This condition is fundamental for circle detection methods, such as the Hough
Transform, which rely on verifying whether a given set of image points conforms to
the geometric constraints of a circle.

$$\rho^2 = (x - u)^2 + (y - v)^2. \tag{8.10}$$

Based on the requirement established by Eq. 8.10, implementing the Hough Transform for circle detection necessitates the use of a three-dimensional accumulator array, denoted as $\mathrm{MRAcc}(u, v, \rho)$. This structure allows the algorithm to account for the three parameters that define a circle: the center coordinates (u, v) and the radius ρ. The overall logic of the circle detection algorithm closely follows the framework used for line detection; however, it incorporates key modifications to accommodate the circular model. Specifically, the algorithm evaluates whether each edge point in the image satisfies the condition of lying on a circle with various centers and radii, as specified by Eq. 8.10. For each valid combination, the corresponding accumulator cell is incremented, allowing the algorithm to identify the most probable circle candidates through local maxima in the accumulator space. The detailed procedure is presented in Algorithm 8.2, outlining each step required for robust and efficient circle detection using the Hough Transform.

Algorithm 8.2 Procedure for circle detection via Hough Transform

	Hough Transform Circle Detector $(I(x, y))$
1:	$0 \rightarrow MRAcc(\theta, r)$
2:	*for* all coordinates of the image $I(x, y)do$
3:	*if* $(I(x, y)$ is an edge$)then$
4:	*forallvalues*$(u, v)do$
5:	$\rho^2 = (x - u)^2 + (y - v)^2$

(continued)

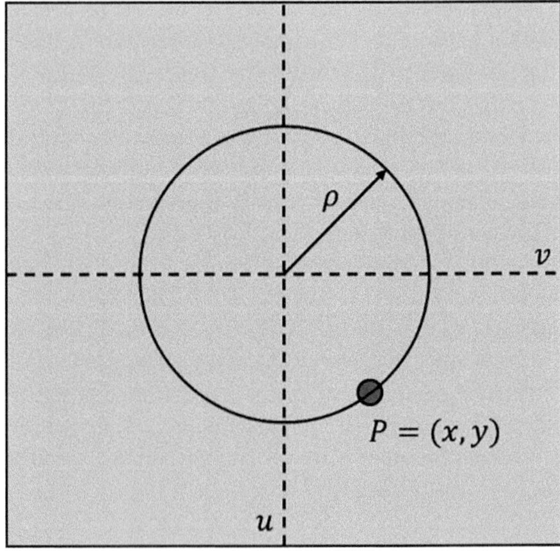

Fig. 8.10 Geometric parameters of a circle

(continued)

6:	$MRAcc(u, v, \rho)$ is incremented
7:	Finally, the $MRAcc(\theta, r)$ registers whose values are maximum are found

As illustrated in Algorithm 8.2, the procedure for detecting circles using the Hough Transform closely mirrors the approach used for line detection, with necessary adaptations to accommodate the three parameters that define a circle. The process begins by initializing all cells in the three-dimensional accumulator matrix to zero (statement 1). Next, the binary image $I(x, y)$ is scanned (statement 3), and whenever an edge pixel is encountered (i.e., the pixel has a value of one), the algorithm uses Eq. 8.10 to compute possible values of the radius ρ, while sweeping through all potential center coordinates (u, v) within the image dimensions (m, n) (statement 4). For each valid triplet (u, v, ρ), the corresponding cell in the accumulator matrix is incremented (statement 6), thereby accumulating evidence for circles with those parameters. Once the entire image has been processed, the accumulator contains peaks—local maxima—that indicate the most probable circle parameters present in the image (statement 7). To extract these candidates, a threshold is first applied to eliminate low-value entries, ensuring only significant accumulations remain. Then, a search is conducted within the three-dimensional parameter space, identifying local maxima by examining a defined volumetric neighborhood. These maxima represent the estimated values of u, v, and ρ that define the circles detected in the image $I(x, y)$.

Figure 8.11 presents a sequence of images illustrating the intermediate results generated by Algorithm 8.2 for circle detection using the Hough Transform. Figure 8.11a shows the artificially created original image, specifically designed to demonstrate the effectiveness of the circle detection process. In Fig. 8.11b, the edges of the image are highlighted, obtained through the application of the Canny edge detection algorithm, as discussed in Chap. 6. Figure 8.11c displays the contents of the three-dimensional accumulator array, which represents the votes accumulated for each possible combination of center coordinates (u, v) and radius ρ. In Fig. 8.11d, the brightest or reddest areas indicate the points in parameter space where the accumulator values have significantly increased—these correspond to the most probable parameter sets (u, v, ρ) that define actual circles present in the original image $I(x, y)$. It's worth noting that due to the volumetric representation and rotations applied for visualization, Figs. 8.11c and d appear visually inverted relative to the spatial arrangement seen in Figs. 8.11a and b. This highlights how transformations in parameter space do not always directly correspond to the spatial orientation of the original image but still provide critical information for accurate shape detection.

A more efficient approach to implementing Algorithm 8.2 involves avoiding an exhaustive search across the entire three-dimensional (u, v, ρ) parameter space. Instead, by fixing a specific value of the radius ρ, one can perform the search for local maxima only within the two-dimensional (u, v) plane. This significantly reduces computational complexity and speeds up the detection process. Figure 8.12 illustrates this optimization by showing the (u, v) slice of the accumulator array for a fixed radius of $\rho = 30$. In this representation, the brightest or reddest points indicate

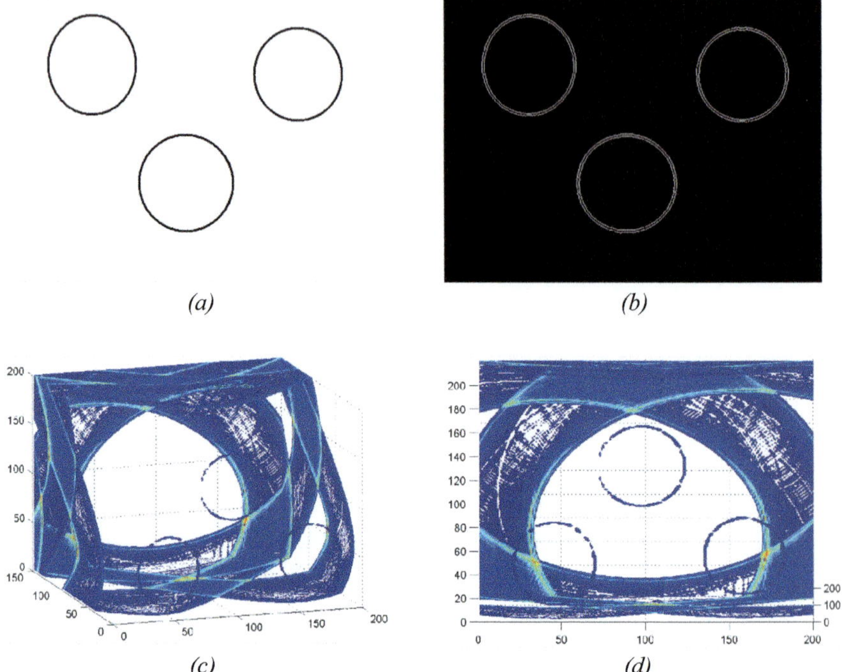

Fig. 8.11 Intermediate results of Algorithm 8.2: **a** the original input image, **b** edge map of the original image obtained using the Canny algorithm, **c** 3D visualization of the accumulator array, and **d** identification of peak values within the accumulator

the highest values in the accumulator, corresponding to the most likely centers of circles with radius 30. These detected centers align with the actual circular structures present in the original image shown in Fig. 8.11a, demonstrating the effectiveness of this targeted search strategy for enhancing performance while maintaining detection accuracy.

The **computational complexity** of the Hough Transform for **circle detection** is significantly higher than that of its line detection counterpart due to the increase in dimensionality of the parameter space. While line detection typically involves a **2D accumulator array** (e.g., slope and intercept or angle and distance), circle detection requires a **3D accumulator** to account for the three defining parameters of a circle: the center coordinates (u, v) and the radius ρ.

This means that for an image of size $M \times N$, and a range of possible radii R, the size of the accumulator space becomes $O(M \cdot N \cdot R)$. For each edge pixel detected in the image, the algorithm must iterate over all possible radius values

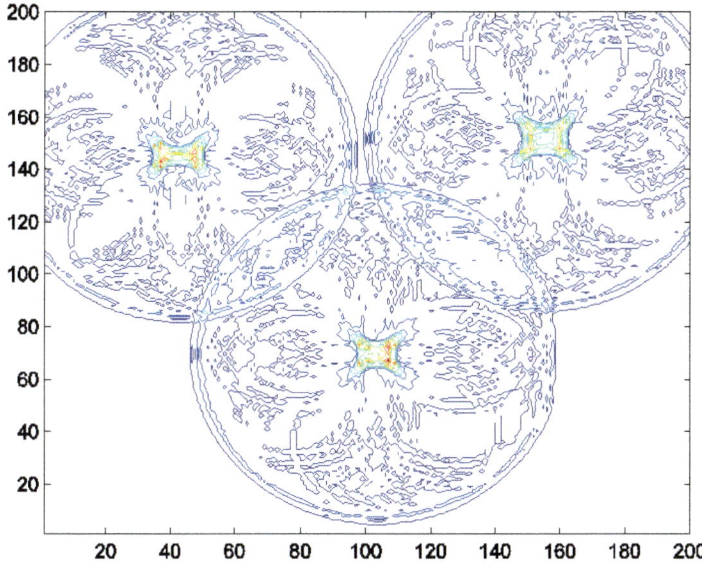

Fig. 8.12 Plane (u, v) of the accumulator array with a fixed radius $\rho = 30$

and compute potential centers that satisfy the circle equation. This results in a total time complexity of $O(E \cdot R)$, where E is the number of edge pixels in the image and R is the number of discrete radius values considered.

As a result, the Hough Transform for circle detection can be computation-ally intensive and memory demanding, especially when high resolution, large images, or a wide range of radii are involved. To mitigate this, optimizations are often employed, such as fixing the radius in advance, limiting the param-eter space, using more efficient accumulator updates, or applying probabilistic variants of the transform.

Despite its computational cost, the method remains popular because of its robustness to noise, occlusion, and partial shapes, making it suitable for a variety of practical applications where accuracy is prioritized over speed. However, in real-time or resource-constrained environments, faster alternatives or hardware acceleration may be necessary.

8.6 Hough Transform for Circle Detector Implemented in Python

Hough Transform For Circle Detector

#CODE 8.2

```python
# The libraries necessary for image processing are loaded.

import cv2 as cv
```

```python
import numpy as np

import matplotlib.pyplot as plt

from google.colab.patches import cv2_imshow
```

```python
img = cv.imread('Image8.1.jpg')

imgGray = cv.cvtColor(img,cv.COLOR_BGR2GRAY)

cv_imshow(imgGray)
```

img2Blur = cv.GaussianBlur(img2Gray,ksize=(9,9),sigmaX=1.5)

cv_imshow(img2Blur)

circles =
cv.HoughCircles(img2Blur,method=cv.HOUGH_GRADIENT,dp=1,minDist=50,param1=20
0,param2=50,minRadius=0,maxRadius=0)

image: An 8-bit, single-channel grayscale input image.

circles: Output vector containing the detected circles, formatted as cv.CV_32FC3. Each circle is represented by a floating-point triplet: $(x,y,radius)$.

method: The detection method used. Refer to cv.HoughModes. Currently, only the HOUGH_GRADIENT method is supported.

dp: The inverse ratio of the accumulator resolution to the input image resolution. For example, dp = 1 means the accumulator and the image have the same resolution; dp = 2 means the accumulator's width and height are half that of the image.

minDist: The minimum allowable distance between the centers of detected circles. A very small value may result in multiple nearby detections for a single circle, while a very large value could cause some circles to be missed.

param1: The first parameter specific to the detection method. For HOUGH_GRADIENT, this is the upper threshold for the Canny edge detector; the lower threshold is automatically set to half of this value.

param2: The second parameter specific to the method. For HOUGH_GRADIENT, it specifies the accumulator threshold used to identify circle centers. A lower value increases sensitivity but may produce more false positives. Detected circles with higher accumulator scores are returned first.

minRadius: The minimum allowable radius of detected circles.

maxRadius: The maximum allowable radius of detected circles.

```python
circles = np.uint16(np.around(circles))

imgCircles = img2.copy()

for i in circles[0,:]:

  # draw the outer circle
```

```
cv.circle(imgCircles,center=(i[0],i[1]),radius=i[2],color=(0,255,0),thickness=2)
```

```
# draw the center of the circle
```

```
cv.circle(imgCircles,center=(i[0],i[1]),radius=2,color=(0,0,255),thickness=5)
```

```
cv_imshow(imgCircles)
```

In addition to the Hough Transform, several other methods are available for detecting circles in images, each suited to different contexts and computational requirements. One common approach is template matching, where circular templates of varying sizes are slid across the image to compare against local regions using similarity measures such as normalized cross-correlation. While simple, this method can be computationally expensive, especially when the circle's scale and orientation vary.

Another technique is based on edge curvature analysis, which examines the local curvature of edge contours. Circles can be identified by tracking continuous curves with consistent curvature values. Although this method is sensitive to noise, it can be effective in well-defined edge maps.

Contour-based methods, such as those using the Douglas–Peucker algorithm, simplify edge contours to detect approximate shapes. Circles can be distinguished by evaluating geometric properties like perimeter-to-area ratios or fitting ellipses and testing for near-circularity.

Gradient vector flow (GVF) and active contours (or "snakes") can also be adapted for circle detection by initializing a circular contour and allowing it to evolve under the influence of internal shape constraints and external image forces until it stabilizes around circular features.

More recently, deep learning-based methods have emerged, using convolutional neural networks (CNNs) to learn and recognize circular shapes directly from labeled training data. These models can detect circles under challenging conditions, such as occlusion or varying lighting, and are particularly useful in real-time applications.

Each of these methods has trade-offs in terms of accuracy, computational cost, robustness to noise, and flexibility, and the choice of method often depends on the specific requirements and constraints of the task at hand.

References

1. Mukhopadhyay, P., & Chaudhuri, B. B. (2015). A survey of Hough transform. *Pattern Recognition, 48*(3), 993–1010.
2. Dahyot, R. (2008). Statistical hough transform. *IEEE Transactions on pattern analysis and machine intelligence, 31*(8), 1502–1509.
3. Matarneh, S., Elghaish, F., Al-Ghraibah, A., Abdellatef, E., & Edwards, D. J. (2025). An automatic image processing based on Hough transform algorithm for pavement crack detection and classification. *Smart and Sustainable Built Environment, 14*(1), 1–22.
4. Huang, Q., & Liu, J. (2021). Practical limitations of lane detection algorithm based on Hough transform in challenging scenarios. *International Journal of Advanced Robotic Systems, 18*(2), 17298814211008752.
5. Huang, A. C., Yuan, C., Meng, S. H., & Huang, T. J. (2023). Design of fatigue driving behavior detection based on circle hough transform. *Big data, 11*(1), 1–17.

Chapter 9
Image Segmentation

This chapter provides a detailed exploration of segmentation one of the most essential processes in computer vision—dividing an image into meaningful regions to facilitate tasks such as object detection, recognition, and scene understanding. It begins by defining segmentation as the classification of pixels into homogeneous regions based on attributes like intensity, color, or texture, distinguishing objects from the background. The chapter first introduces threshold-based segmentation, explaining how pixels are divided according to intensity values and discussing both ideal cases with well-separated distributions and real cases with overlapping ones. Techniques such as median filtering are presented to improve segmentation quality by reducing noise. The concept of an optimal threshold is then developed through statistical reasoning, leading to a mathematical derivation and Python implementation of Otsu's algorithm, which determines the threshold that minimizes intra-class variance or maximizes inter-class separation. Beyond global thresholding, the chapter introduces region-growing segmentation, a contextual technique that iteratively groups neighboring pixels based on intensity similarity, starting from a seed pixel. The mathematical formulation, algorithmic steps, and complete Python implementation of the region-growing method are described in detail, demonstrating adaptive segmentation of complex images. Through theoretical exposition, mathematical modeling, and practical coding examples, the chapter offers a comprehensive understanding of both statistical and spatial approaches to image segmentation, establishing a foundation for advanced analysis and object-based interpretation.

Image segmentation is a fundamental and active area of research within the field of computer vision, driven by its broad range of applications, including object detection, recognition, classification, and scene interpretation. It serves as a crucial first step in many image analysis and pattern recognition tasks [1].

At its core, image segmentation involves dividing an image into distinct, meaningful regions. These regions may correspond to complete objects, parts of objects, or background areas, grouped based on visual similarity in attributes like intensity, color, texture, or spatial relationships.

© The Author(s), under exclusive license to Springer Nature Switzerland AG 2026
E. Cuevas et al., *Image Processing with Python*, Signals and Communication
Technology, https://doi.org/10.1007/978-3-032-13285-7_9

A variety of methods and algorithms have been developed to perform segmentation. The selection of a suitable technique depends on the specific characteristics of the application, the type of image data, and the features being analyzed. Some methods emphasize edge detection or region growing, while others rely on statistical properties or machine learning.

A common outcome of the segmentation process is a binary image, where each pixel is assigned one of two possible values (typically 1 or 0) to distinguish the foreground (object of interest) from the background. However, real-world images often present complex scenarios with overlapping or ambiguous boundaries, making this simplification only a starting point.

In this chapter, we will examine the most widely used segmentation techniques, exploring both their theoretical foundations and practical applications. Special attention will be given to the generation and analysis of binary images, including the isolation and structural description of individual objects. A connected component in a binary image is a group of adjacent pixels (connected by a neighborhood rule) that share the same classification and are surrounded by background pixels.

9.1   Segmentation

Segmentation is the process of dividing an image into multiple regions such that the pixels within each region share similar characteristics, while pixels in different regions differ significantly according to a given criterion. The main goal is to distinguish objects of interest from the rest of the scene, typically referred to as the background [2].

In its simplest form, segmentation results in a binary classification of pixels: those belonging to the object and those corresponding to the background. This process is often referred to as binarization, and it plays a key role in preparing images for higher-level tasks such as object classification or scene interpretation.

The most commonly used criterion for determining similarity in segmentation is pixel intensity. However, for more complex or detailed applications, additional features such as color, texture, or gradient information may be considered to enhance accuracy.

Segmentation methods can be broadly categorized based on the features they utilize and the type of algorithmic approach they employ. The main categories are contextual and non-contextual methods.

Non-contextual methods consider only the global characteristics of pixels, ignoring their spatial relationships. One of the most well-known non-contextual methods is thresholding, where each pixel is classified based on whether its intensity exceeds a predefined threshold value. Contextual methods incorporate additional spatial or structural information, taking into account the relationships between neighboring pixels. For example, contextual techniques may group pixels not only by intensity similarity but also by proximity or directional coherence, such as consistent gradient orientation.

By combining feature analysis with contextual relationships, segmentation methods can better capture the structure and coherence of real-world objects. In the following sections, we will delve into both non-contextual and contextual techniques, illustrating their principles and applications through practical examples.

9.2 Thresholding

Thresholding is a basic yet powerful image segmentation technique. It operates under the assumption that the objects of interest in an image are composed of pixels with relatively homogeneous intensity values, which are distinguishable from those of the background or other objects [3].

The process consists of comparing the intensity of each pixel with a predefined threshold value. If the pixel's intensity is greater than the threshold, the pixel is classified as belonging to one category (typically the object); otherwise, it is assigned to the other category (commonly the background). As a result, thresholding produces a binary image where the segmentation decision is made independently for each pixel based on intensity.

However, the quality of the segmentation is highly dependent on the choice of the threshold. In an ideal scenario, the intensity distributions of the different objects in the image are well separated, with no overlap. In such a case, selecting an appropriate threshold is straightforward.

9.2.1 Ideal Case: Separated Distributions

Figure 9.1a presents a synthetic image composed of two distinct square objects: object A and object B. Object A contains pixels with intensities drawn from a Gaussian distribution with a mean of 60 and a standard deviation of 10. Object B, on the other hand, is composed of pixels with a Gaussian distribution centered at 190, also with a standard deviation of 10.

Figure 9.1b shows the corresponding intensity histogram of the image. The two Gaussian distributions appear clearly separated, making it easy to select a threshold value (128, in this case) that cleanly divides both objects. Pixels with intensity greater than 128 are classified as belonging to object B, and the rest to object A.

The result of applying thresholding with a value of 128 to this image is shown in Fig. 9.2, where a clean binary segmentation is obtained [4].

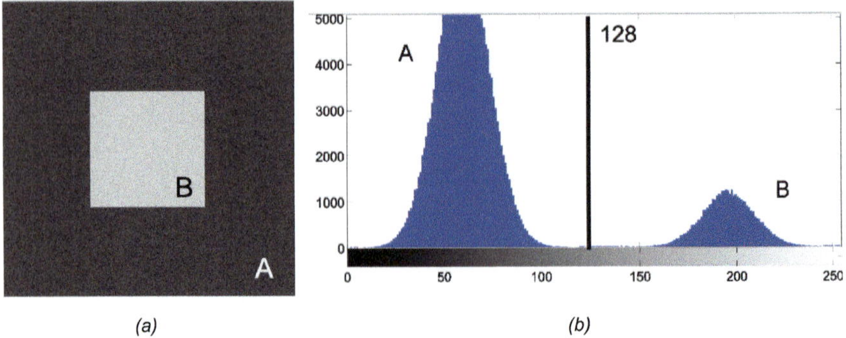

Fig. 9.1 Distribution of the intensities of the pixels that make up two different objects in an image.
a The two objects and **b** their distribution in the histogram

Fig. 9.2 Segmentation
result of Fig. 5.1a
considering a threshold value
of 128

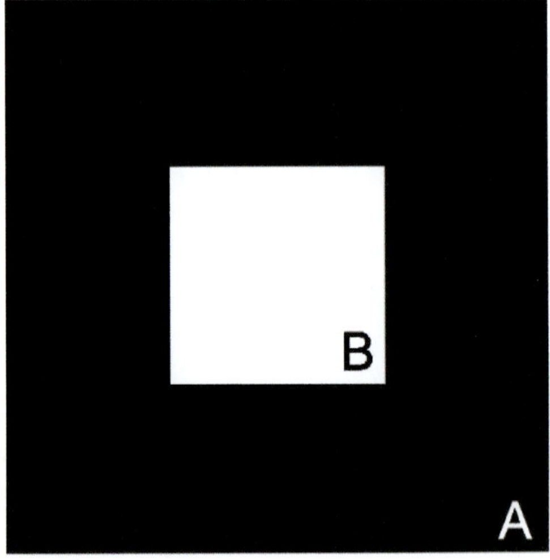

9.2.2   Real Case: Overlapping Distributions

In practice, pixel intensity distributions of different objects frequently overlap, due
to noise, lighting variations, texture, or other real-world complexities. This overlap
introduces uncertainty into the segmentation, making it difficult to decide which
pixels belong to which object.

Figure 9.3a presents an image with two square objects whose intensity distributions now overlap significantly. Figure 9.3b shows the resulting histogram, where
the two Gaussian peaks are no longer clearly separated [5].

Applying thresholding to this image using a value of 150, roughly midway
between the two modes, results in imperfect segmentation, as shown in Fig. 9.4.

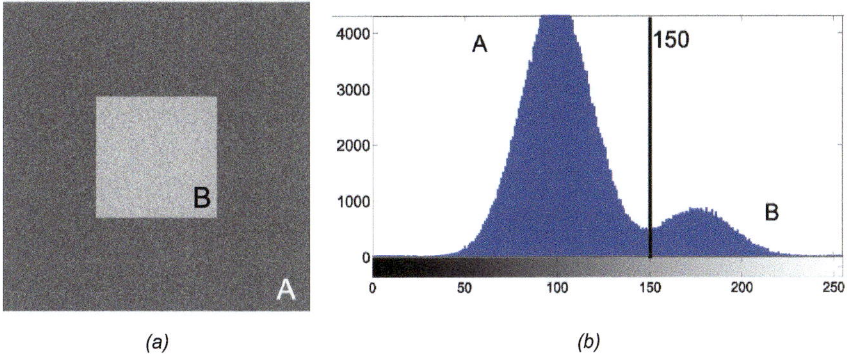

Fig. 9.3 Objects overlapping their distributions. **a** The two objects and **b** their distribution in the histogram

Fig. 9.4 Segmentation result of Fig. 5.3a considering a threshold value of 128

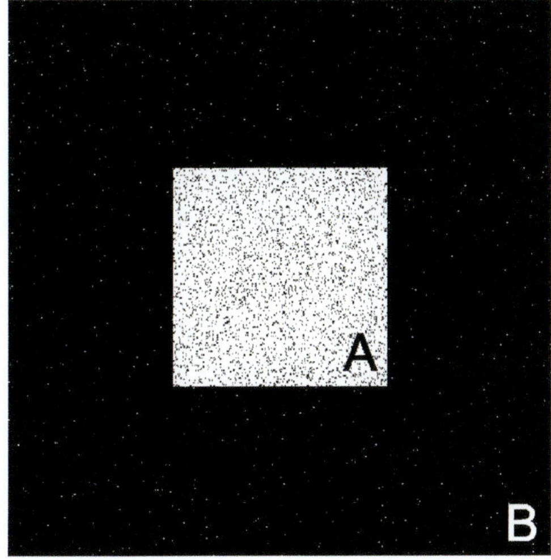

Some pixels within the overlapping region are incorrectly classified, leading to segmentation errors.

9.2.3 *Improving Segmentation: Median Filter Preprocessing*

To address the problem of overlapping distributions, a common solution is to apply a median filter to the image before thresholding. The median filter is a nonlinear filter that smooths the image while preserving edges. It helps reduce noise and eliminate small intensity variations that cause distribution overlap [6].

Figure 9.5a shows the original image from Figs. 9.3a and 9.5b shows the result after applying a median filter with a 3×3 neighborhood. As seen in Fig. 9.5c, the histogram of the filtered image exhibits reduced overlap between the intensity distributions, making it easier to identify an appropriate threshold.

Using a new threshold value of 145, binarization of the filtered image yields a much cleaner segmentation, as shown in Fig. 9.6. The improvement demonstrates the utility of preprocessing in threshold-based segmentation methods.

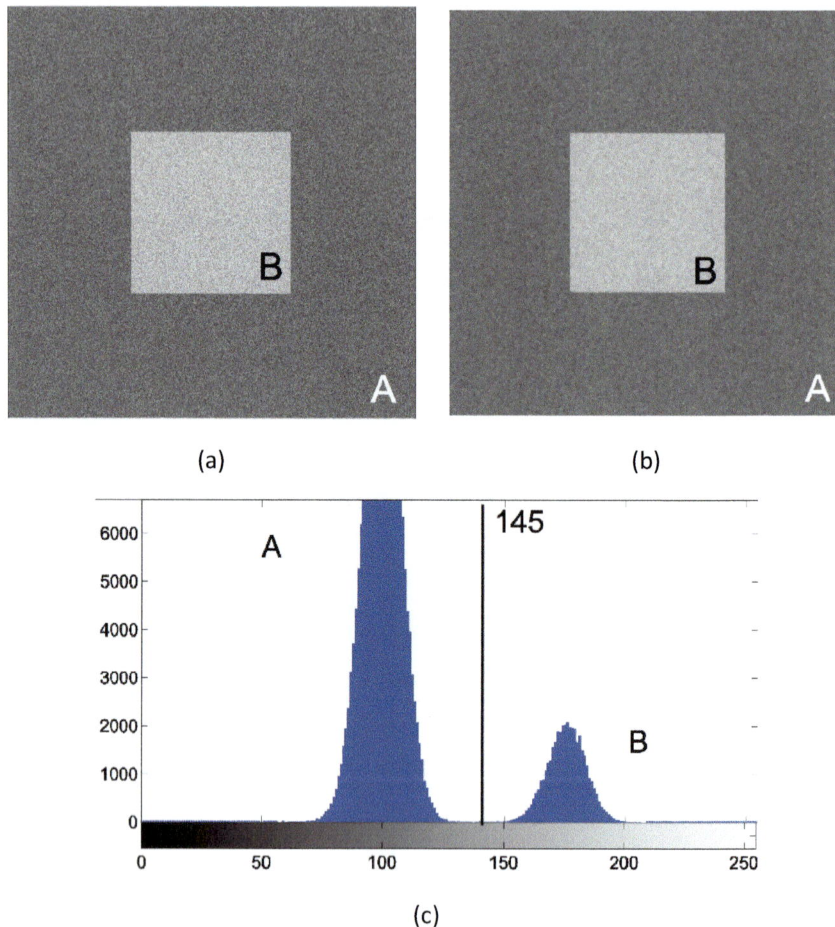

Fig. 9.5 Effect of using the median filter on intensity distributions. **a** The original image, **b** the image obtained by filter processing, and **c** the resulting distribution

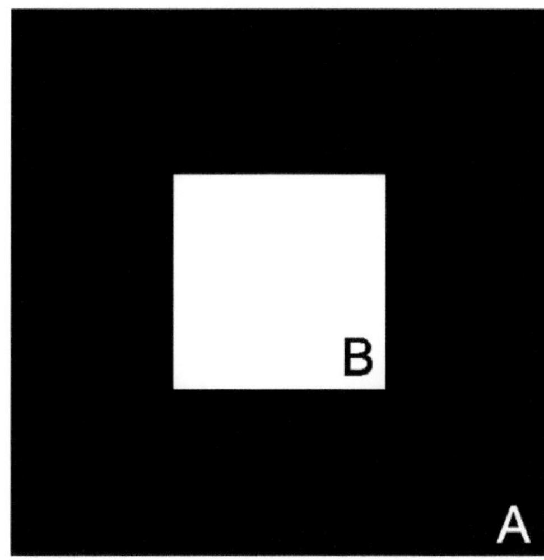

Fig. 9.6 Segmentation result of Fig. 5.b (a) considering as threshold the value of 145

9.3 Calculating the Optimum Threshold

As seen in the previous section, overlapping intensity distributions present a significant challenge for threshold-based segmentation. Although median filtering and other preprocessing techniques can help reduce noise and smooth intensity values, they do not always eliminate the overlap between distributions, especially when the object and background intensities are inherently similar [7].

In such cases, choosing an appropriate threshold becomes critical. The goal is to find the optimal threshold that minimizes the number of misclassified pixels, that is, pixels from object A incorrectly classified as part of object B, and vice versa.

The optimal threshold corresponds to the intersection point of the two distributions in the histogram. At this point, the probability of error on either side is balanced. In other words, the threshold is set where the probability of misclassifying a pixel from object A as belonging to object B is approximately equal to the opposite case.

Figure 9.7 illustrates this concept graphically for a hypothetical example. It presents two overlapping Gaussian distributions, each representing the intensity profile of a different object (A and B). The shaded region T represents the area of overlap, where classification is uncertain. The intersection point between the distributions serves as the best estimate for the threshold, as it minimizes overall classification error.

In practice, determining this optimal threshold analytically requires knowledge of the underlying probability distributions of the objects, which is rarely available in real images. Consequently, several iterative algorithms have been proposed to estimate the best threshold value based on intensity statistics.

Most of these methods rely on the following principle: For a given candidate threshold, pixels are grouped into two classes. The threshold is adjusted iteratively so that each pixel is assigned to the class whose mean intensity is closer to that of the

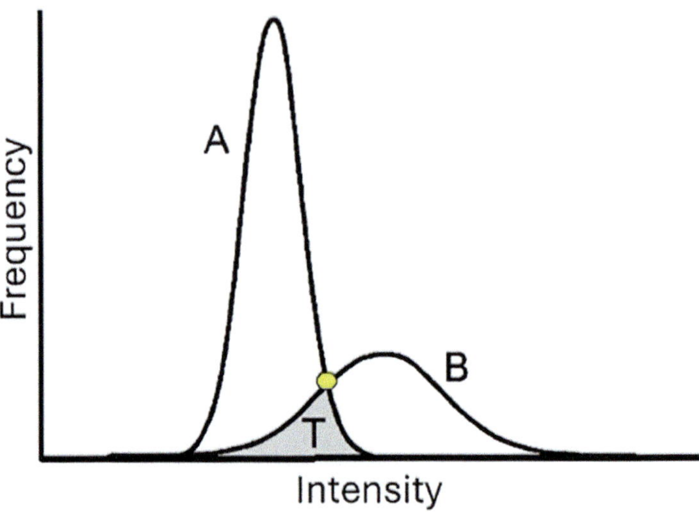

Fig. 9.7 Optimal threshold determined by the intersection of two overlapping distributions. The shaded area T represents the region of misclassification. The intersection point marks the minimum error boundary between objects A and B

pixel, thereby minimizing intra-class variance or maximizing inter-class separation. One widely used method that follows this principle is Otsu method, which will be discussed in detail in the next section.

9.4 Otsu Algorithm

As introduced in the previous section, determining the optimal threshold value is crucial when handling overlapping intensity distributions. Among the various methods proposed for this purpose, the Otsu algorithm stands out as one of the most widely used and practical techniques in image segmentation.

The Otsu method, proposed by Nobuyuki Otsu in 1979, is a non-parametric, unsupervised technique that automatically determines the optimal threshold value for binarizing an image into two classes: foreground and background. It is particularly well-suited for cases where the image histogram shows a bimodal distribution, but it can also offer reasonable performance in more complex scenarios.

Otsu's method works by analyzing the histogram of gray levels in an image and finding the threshold k that minimizes the intra-class variance (i.e., the variance within each of the two groups of pixels formed by the threshold) or, equivalently, maximizes the inter-class variance (i.e., the separation between the two groups).

Given a grayscale image I of dimensions MxN, with gray levels ranging from 0 to $L - 1$, the histogram of the image describes the frequency of each intensity level.

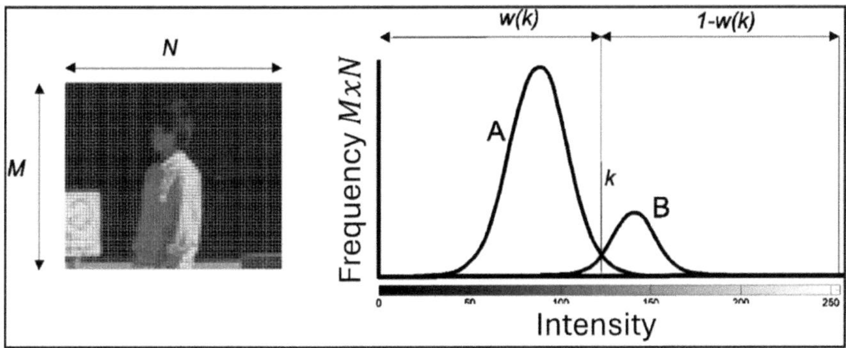

Fig. 9.8   Histogram of an image with two overlapping distributions, divided by a candidate threshold k

The method assumes that the histogram can be divided into two groups based on a candidate threshold value k: object A pixels with intensity values from 0 to k, and object B pixels with intensity values from $k + 1$ to $L - 1$.

Otsu's algorithm evaluates each possible threshold $k \in [1, L - 2]$ and computes the weighted intra-class variance of the two groups. The optimal threshold is the one that minimizes this measure, resulting in the best possible separation of the two-pixel classes.

Figure 9.8 illustrates the general idea: a bimodal histogram where the intensity values are divided into two groups by a candidate threshold k along with the associated statistical quantities needed for the calculation.

9.4.1   *Mathematical Formulation*

The Otsu method begins by computing the probability distribution of pixel intensities in the image:

$$P(A) = \sum_{i=0}^{k} h(i) = w_A(k) = w(k),$$

$$P(A) = \sum_{i=k+1}^{L-1} h(i) = w_B(k) = 1 - w(k), \tag{9.1}$$

where $h(i)$ corresponds to the frequency of pixels that exhibit an intensity value of i in the image. The mean intensity values for each class are given by:

$$\mu_A(k) = \frac{1}{W_A(k)} \sum_{i=0}^{k} i \cdot h(i),$$

$$\mu_B(k) = \frac{1}{W_B(k)} \sum_{i=k+1}^{L-1} i \cdot h(i). \tag{9.2}$$

Based on these values, the variances for each class can be derived as follows:

$$\sigma_A^2(k) = \frac{1}{W_A(k)} \sum_{i=0}^{k} (i - \mu_A(k))^2 \cdot h(i),$$

$$\sigma_B^2(k) = \frac{1}{W_B(k)} \sum_{i=k+1}^{L-1} (i - \mu_B(k))^2 \cdot h(i). \tag{9.3}$$

Using these parameters, we can determinate the between-class variance, which is a measure of class separability:

$$\sigma_D^2 = W_A(k) \cdot \sigma_A^2(k) + W_B(k) \cdot \sigma_B^2(k). \tag{9.4}$$

Building on the concepts previously defined, the Otsu algorithm aims to determine the optimal threshold value k that minimizes the intra-class variance, denoted as σ_D^2, as shown in Eq. 9.4. The underlying intuition is straightforward: by minimizing the variability within each of the segmented regions the distinction between them becomes clearer, thus reducing their statistical overlap. A lower intra-class variance implies that the pixel intensities in each group are more tightly clustered around their respective means, which enhances the separability between the classes.

A simple implementation of the algorithm would involve evaluating all possible threshold values from 0 to $L - 1$ calculating the intra-class variance for each case and selecting the threshold that yields the minimum value of σ_D^2. While this brute-force approach is conceptually simple and guarantees correctness, it can become computationally expensive for large images or when applied in real-time scenarios.

To address this issue, the Otsu method takes advantage of a recursive formulation that enables more efficient computation. Instead of recalculating the required statistical quantities from scratch for each threshold candidate, one can update them incrementally. This recursive optimization significantly reduces the computational load without compromising accuracy. However, this requires computing several components, including the total variance and the between-class variance, which are defined as follows.

The total variance σ_T^2, which is independent of the threshold k, is computed based on the entire intensity distribution of the image:

$$\sigma_T^2 = \sum_{i=0}^{L-1} (i - \mu_T)^2 \cdot h(i), \tag{9.5}$$

where μ_T is the global mean intensity of the image, calculated by:

$$\mu_T = \sum_{i=0}^{L-1} i \cdot h(i). \tag{9.6}$$

Using these definitions, the between-class variance σ_E^2, which reflects the degree of separation between the two classes, can be calculated by subtracting the intra-class variance from the total variance. This expression can also be reformulated in different but equivalent ways that are more computationally convenient:

$$\begin{aligned}
\sigma_E^2 &= \sigma_T^2 - \sigma_D^2 \\
&= w(k) \cdot (\mu_A(k) - \mu_T)^2 - (1 - w(k)) \cdot (\mu_B(k) - \mu_T)^2 \\
&= w(k) \cdot (1 - w(k)) \cdot (\mu_A(k) - \mu_B(k))^2.
\end{aligned} \tag{9.7}$$

Since the total variance σ_T^2 remains constant regardless of the threshold, the optimization problem becomes a matter of maximizing the between-class variance σ_E^2. Maximizing σ_E^2 is computationally advantageous because this function is smooth and unimodal (it has a single maximum) which guarantees that a simple incremental evaluation across all k values will reliably yield the optimal threshold.

To further enhance efficiency, the Otsu method uses the following recursive equations to update class probabilities and means incrementally:

$$\begin{aligned}
w(k + 1) &= w(k) + h(k + 1) \\
\mu_A(k + 1) &= \frac{w(k) \cdot \mu_A(k) + (k + 1) \cdot h(k + 1)}{w(k + 1)} \\
\mu_B(k + 1) &= \frac{\mu_T - w(k + 1) \cdot \mu_A(k + 1)}{1 - w(k + 1)}.
\end{aligned} \tag{9.8}$$

The recursive update begins with the initial values: $w(1) = h(1)$ and $\mu_A(0) = 0$. As the threshold k progresses from 0 to $L-1$, the algorithm updates the class statistics using the expressions above and checks whether the new σ_E^2 value is greater than the previous maximum. The threshold associated with the highest σ_E^2 is selected as the optimal threshold.

This recursive approach enables an efficient and elegant implementation of the Otsu algorithm. Code 9.1, written in Python, demonstrates how this method can be implemented in practice. It first converts a color image into grayscale and then applies the Otsu thresholding process by following the mathematical steps described from Eqs. 9.1 through 9.8. The results of applying this method are shown in Fig. 9.9, which presents both the original image and its binarized version obtained using the optimal threshold.

(a)   (b)

Fig. 9.9 Otsu method results for binarization produced by code 9.1. **a** Original image and **b** result image

Otsu Method for Binarization of Distribution of Two Overlapping Classes

 **#CODE 9.1**

```python
# The libraries necessary for image processing are loaded.
import cv2 as cv
import numpy as np
import matplotlib.pyplot as plt
from google.colab.patches import cv2_imshow
import sys
from skimage.filters import threshold_otsu
```

```python
img = cv.imread('Image9.1.jpg')
img = cv.cvtColor(img,cv.COLOR_BGR2GRAY)
cv_imshow(img)
```

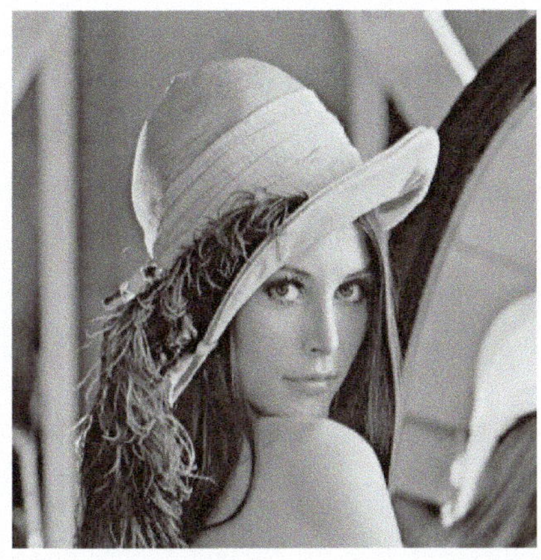

```python
h = cv.calcHist([img],channels=[0],mask=None,histSize=[256],ranges=[0,255])
plt.plot(h)
plt.show()
```

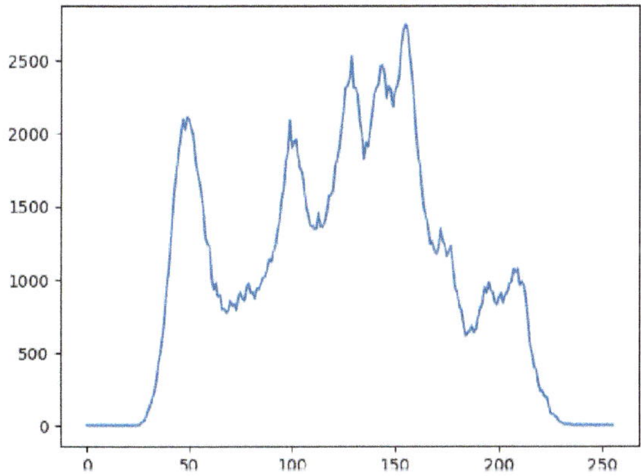

```python
# Otsu's Algorithm
i = np.linspace(0,255,256)
```

```python
min = np.inf
var_vec = np.zeros(256)
for k in range(256):
  wA = np.sum(h[0:k])
  wB = np.sum(h[k+1:255])
  # Compute mean for class A
  mA = np.dot(i[0:k], h[0:k]) / (wA + sys.float_info.epsilon)
  # Compute mean for class B
  mB = np.dot(i[k+1:255], h[k+1:255]) / (wB + sys.float_info.epsilon)
  # Compute variance for classes A and B
  vA = np.dot((i[0:k] - mA)**2, h[0:k]) / (wA + sys.float_info.epsilon)
  vB = np.dot((i[k+1:255] - mB)**2, h[k+1:255]) / (wB + sys.float_info.epsilon)
  # Compute within-class variance
  vC = (wA * vA) + (wB * vB)
  var_vec[k] = vC[0]
  # Check if a new minimum variance has been found
  # If so, update the threshold
  if (vC[0] <= min ):
    th = k
    min = vC[0]


print('threshold=',th)
```

threshold= 118

```python
# Plot of the variances
plt.plot(var_vec)
plt.scatter(th,min,c='r')
plt.xlabel('Threshold k')
plt.ylabel('Within-class Variance')
plt.show()
```

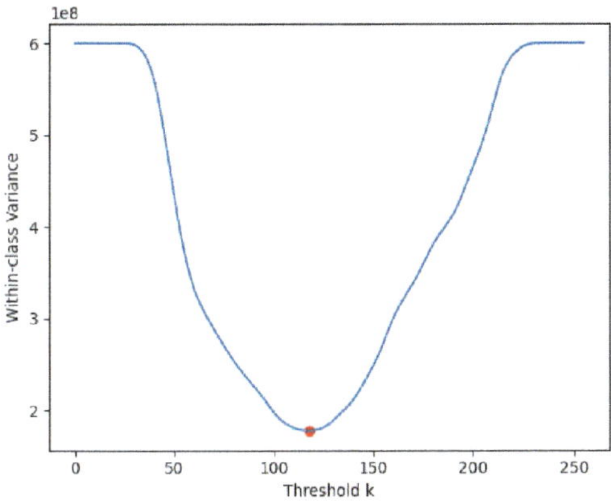

```python
# The image is binarized using the optimal threshold
imgThres_otsu = np.uint8((img >= th).astype(int)*255)
cv_imshow(imgThres_otsu)
```

```
h = cv.calcHist([img],channels=[0],mask=None,histSize=[256],ranges=[0,255])
```

```
plt.plot(h)
plt.vlines(th,ymin=0,ymax=2500,colors='r')
plt.show()
```

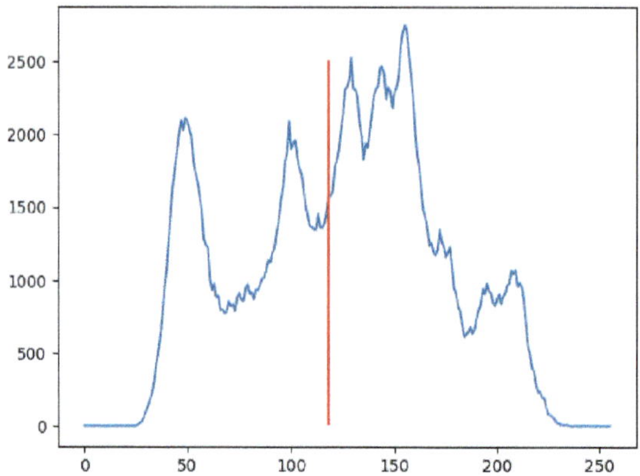

9.5   Region-Growing Segmentation

The region-growing technique is a classic and intuitive image segmentation approach that builds regions by progressively aggregating pixels that satisfy a defined similarity criterion. The core idea is to start from a predefined location (seed) in the image and iteratively include adjacent pixels that share similar attributes.

The process begins by selecting a seed pixel $s(x, y)$,, which serves as the initial reference point for region expansion. The choice of this pixel is crucial: it should ideally lie within the region of interest that is to be segmented. Starting from the seed, the algorithm examines its neighboring pixels (typically using a 4-neighborhood or 8-neighborhood connectivity model) to determine whether each neighboring pixel is similar enough to be considered part of the same region.

The similarity criterion used to compare pixels can vary depending on the application and image characteristics. Common criteria include grayscale intensity, color values, texture features, or even more complex statistical measures. In many implementations, especially in grayscale images, the intensity difference between the seed and candidate pixel is the principal factor in determining similarity.

If a neighboring pixel meets the similarity condition, it is added to the growing region. The process continues recursively or iteratively by examining the neighbors of all the newly added pixels. If a neighbor does not meet the similarity requirement, it is not discarded outright, instead, it may be placed in a temporary list L of unassigned pixels. These pixels can be revisited later as potential seeds for other regions or for re-evaluation in the context of other neighboring regions.

This iterative expansion continues until no further pixels meet the similarity criteria, thereby finalizing the current segmented region. The algorithm can then proceed to select another seed from the remaining unsegmented pixels, repeating the process until the entire image is segmented.

To provide a comprehensive computational understanding of the region-growing approach, the following section outlines the algorithm in a detailed, step-by-step format, allowing for practical implementation and adaptation to various types of image data and segmentation goals.

9.5.1 Initial Pixel Selection

The success of the region-growing segmentation process strongly depends on the correct selection of the initial pixel, also known as the seed pixel $s(x, y)$. This pixel acts as the starting point for building the segmented region and should ideally reside within the boundaries of the region of interest. If the seed lies outside or near the edges of the desired region, the algorithm may produce inaccurate or incomplete results.

In many cases, selecting the seed pixel purely at random or without guidance may lead to inefficient or incorrect segmentation. However, prior knowledge about the image (such as spatial location, expected object characteristics, or contextual cues) can significantly improve the selection process. For instance, choosing a pixel that is centrally located within a known object can reduce computation and improve segmentation accuracy by limiting unnecessary evaluations.

One practical and effective strategy is to select the seed interactively. In this approach, a user visually inspects the image and manually clicks or specifies the location of the seed. This interactive method ensures that the seed lies within the correct region and gives users control to adjust or refine the segmentation as needed, especially in cases where automatic seed selection is unreliable due to noise or low contrast.

Interactive selection is particularly useful in medical imaging, remote sensing, or microscopy, where domain knowledge helps to target specific structures. Nonetheless, in automated systems, seed selection can also be based on preprocessing steps, such as detecting high-confidence regions based on intensity thresholds, edge maps, or blob detection algorithms.

Ultimately, careful seed selection not only determines where the region-growing process begins but also significantly influences the speed, accuracy, and completeness of the segmentation result.

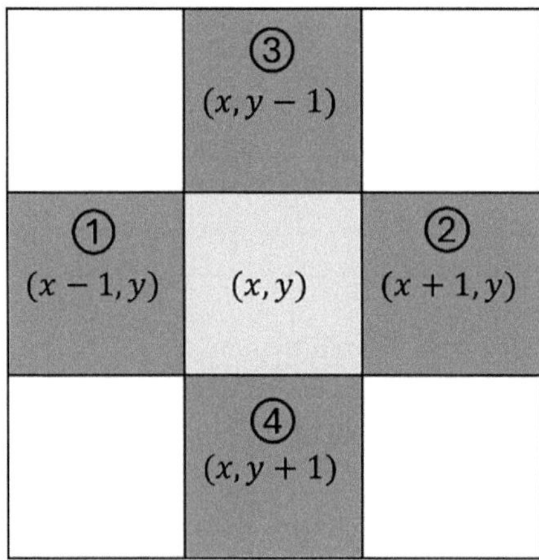

Fig. 9.10 Matching process between the N Matrix and the neighborhood

9.5.2 *Local Search*

Once the seed pixel $s(x, y)$ is defined, the region-growing algorithm initiates a recursive local search around it, aimed at progressively incorporating neighboring pixels that exhibit similar characteristics based on a defined similarity criterion. The process takes place within a local neighborhood V, usually limited to the four nearest neighbors (commonly referred to as a 4-connected neighborhood).

The neighborhood V can be expressed through a displacement matrix, typically defined as:

$V = [-10; 10; 0 - 1; 01]$. Each row in this matrix represents a directional offset (up, down, left, and right), where the two columns correspond to displacements in the horizontal and vertical directions, respectively. This structure allows the algorithm to evaluate adjacent pixels efficiently. Figure 9.10 visually illustrates how this matrix maps onto the local neighborhood.

During the region-growing process, two practical challenges must be addressed to ensure robust segmentation: the handling of image boundaries and the management of search direction. The first issue arises when the neighborhood search reaches the borders of the image, potentially leading to the selection of pixels that fall outside the valid image domain. To prevent this, a validation step is incorporated into the algorithm, verifying that any pixel selected through displacement operations remains within the predefined spatial limits of the image. This ensures that only existing pixels are considered during the segmentation process, thereby avoiding errors or exceptions during execution.

Simultaneously, the second challenge concerns the directional nature of the pixel search. As the algorithm explores the immediate neighborhood around the seed pixel, it may follow a specific direction where similar pixels are continuously found, potentially neglecting other directions in which equally similar pixels exist but are not initially reached. To counter this limitation and ensure a more exhaustive exploration, the algorithm maintains a list L of unselected candidate pixels. This list records both the position and intensity of pixels that were not included in the current iteration but may still satisfy the similarity criteria. When a new pixel pn is added to the region, its intensity is compared against the entries in list L, which may reveal the presence of a distant pixel pd with comparable characteristics, indicating that nearby similar pixels around pn have been exhausted and a redirection of the search is necessary. In such cases, the algorithm retrieves the corresponding pixel information from list L and resumes the segmentation in a new direction, allowing it to capture additional structures that were initially bypassed. As pixels are confirmed as similar and incorporated into the segmented region R, they are removed from list L,, ensuring that all relevant areas are progressively and accurately segmented (Fig. 9.11).

Two essential parameters govern the similarity evaluation during the region-growing segmentation process: the maximum permissible dissimilarity and the similarity of the segmented region. The maximum permissible dissimilarity, denoted as Mp, establishes the threshold beyond which two pixels are no longer considered similar; in other words, it defines the maximum allowable difference in intensity between a candidate pixel and the reference value. If the absolute difference exceeds this threshold, the pixel is excluded from the region. This parameter is predefined as part of the algorithm's configuration and directly influences its sensitivity to intensity variations. On the other hand, the similarity of the segmented region, denoted as Srs, represents the average intensity of the region R currently being formed. Unlike using a fixed-intensity reference, Srs is dynamically updated as new pixels are added to the region, enabling the method to adapt to gradual intensity changes and thereby improving segmentation robustness. This adaptive update is computed iteratively using the following model:

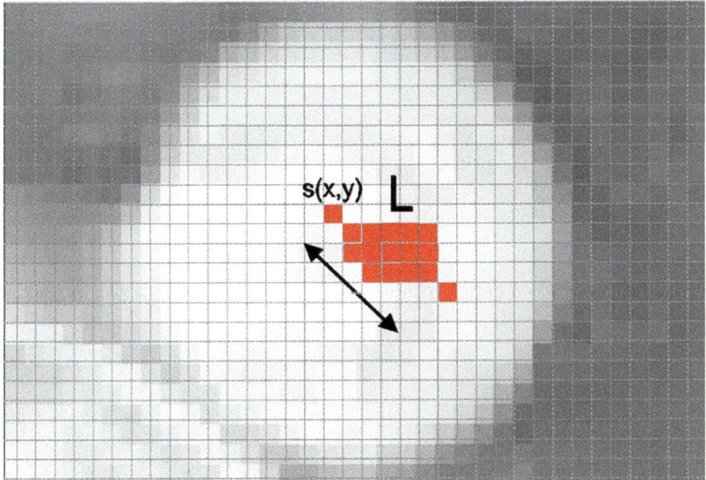

Fig. 9.11 Storage process of the L list in an image

$$\mathrm{Srs}(k+1) = \frac{\mathrm{Srs}(k) \cdot |R| + I(\mathrm{pn})}{|R| + 1}, \tag{9.9}$$

where the value of $\mathrm{Srs}(k+1)$ expresses the updated value of $\mathrm{Srs}(k)$, while $|R|$ represents the number of elements currently constituting the segmented region. Finally, $I(\mathrm{pn})$ defines the intensity of the new pixel pn.

Where $\mathrm{Srs}(k)$ is the current average intensity of the region, $|R|$ is the number of pixels already included in it, and $I(\mathrm{pn})$ denotes the intensity of the new candidate pixel. The result, $\mathrm{Srs}(k+1)$, provides the updated similarity measure that guides the inclusion of subsequent pixels. This formulation ensures that the region can grow adaptively, incorporating pixels with similar but not necessarily identical characteristics, which is crucial for handling natural variations in real-world images.

The implementation of this method is illustrated in Code 9.2, which provides the Python code for region-growing segmentation. Figure 9.12 shows the input grayscale image and the result after applying the region-growing algorithm.

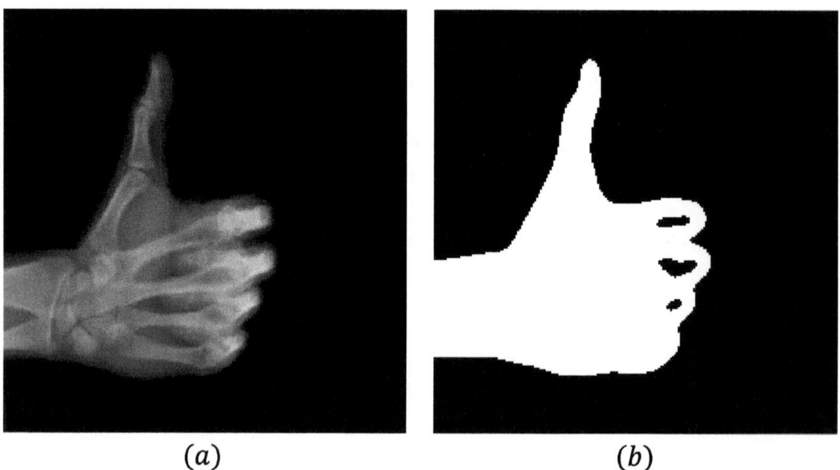

(a)   (b)

Fig. 9.12 Results of the region-growing segmentation method produced by Code 5.2. **a** Original image and **b** result image

Segmentation Method by Region Growth

#CODE 9.2

```python
# The libraries necessary for image processing are loaded.
import cv2 as cv
import numpy as np
import matplotlib.pyplot as plt
from google.colab.patches import cv2_imshow


img = cv.imread('Figure9.12a.jpg',cv.IMREAD_GRAYSCALE)
cv_imshow(img)
```

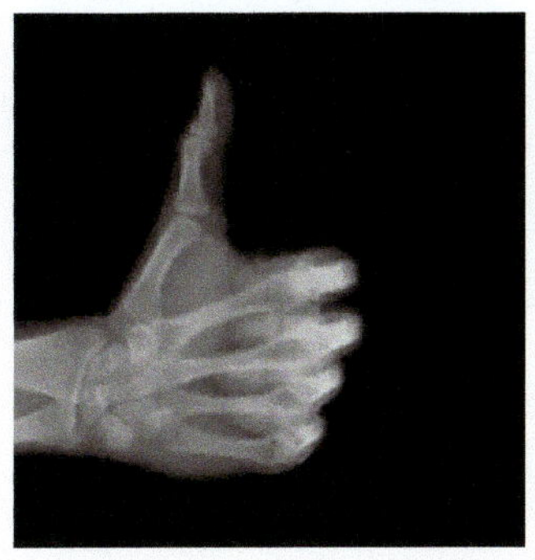

```python
x_start = 150 # @param {"type":"integer"}
y_start = 50 # @param {"type":"integer"}
img2 = img.copy()
cv.circle(img2,center=(y_start,x_start),radius=5,color=(255,0,0),thickness=-1)
cv_imshow(img2)
```

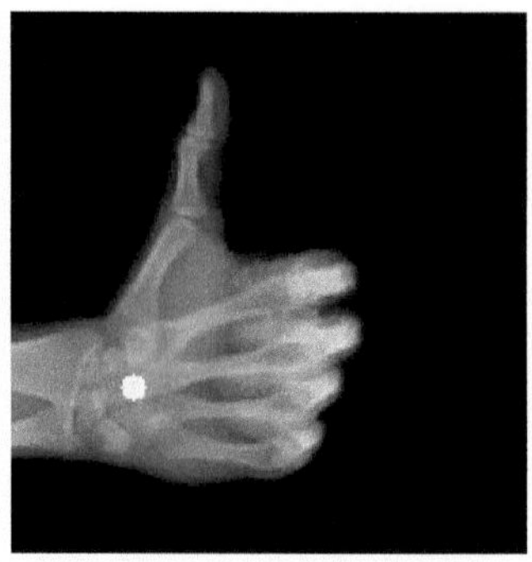

```python
ksize = 9 # @param {"type":"integer"}
imgGauss = cv.blur(img,ksize=(ksize ,ksize))
#imgGauss = cv.GaussianBlur(img,(9,9),0.5)
cv_imshow(imgGauss)
```

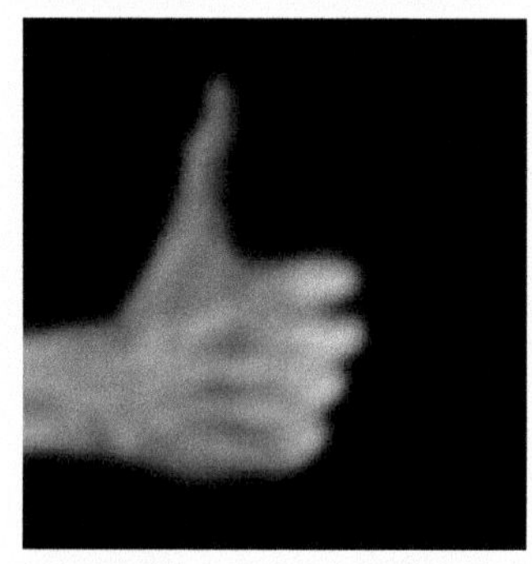

```python
# Parameters are configured

#  Similarity threshold th
th = 0.25 # @param {"type":"number"}
x = x_start
y = y_start
M,N = img.shape
R = 1 #
J = np.zeros((M,N))
V = np.array([[-1,0],[1,0],[0,-1],[0,1]])
pixdist = 0
```

```python
neg_pos = -1
L = np.zeros((img.size,3))
imgNorm = imgGauss / 255
Imean = imgNorm[x,y]


# While the similarity of the selected pixel is less than the minimum similarity
threshold and the number of pixels in the segmented region is less than the size of the
image

while pixdist < th and R < img.size:
  # For each 4-connected neighbor
  for j in range(4):
    # Take a neighboring pixel
    xn = (x + V[j,0]).astype(int)
    yn = (y + V[j,1]).astype(int)
    # Check that the neighboring pixel is within the image boundaries
    ins = (xn >= 0) and (yn >= 0) and (xn < M) and (yn < N)
    # If the neighboring pixel is inside the image and has not been visited
    if ins and J[xn,yn] == 0:
      neg_pos = neg_pos + 1
      # Add the pixel to the list
      L[neg_pos,:] = [xn, yn, imgNorm[xn,yn]]
      # Mark it as visited
      J[xn,yn] = 1
  # Calculate the similarity of each neighboring pixel with the average intensity of the
segmented region
  dist = np.abs(L[0:neg_pos,2] - Imean)
  # Select the pixel with the highest similarity
```

```python
    pixdist = np.min(dist)
    index = np.argmin(dist)
    # Mark the pixel as part of the segmented region with a 2
    J[x,y] = 2
    # Update the average intensity of the segmented region
    R = R + 1
    #Se actualiza la intensidad promedio de la región segmentada
    Imean = (Imean*R + L[index,2]) / (R+1)
    # The selected pixel becomes the new center pixel for the neighborhood
    x = (L[index,0]).astype(int)
    y = (L[index,1]).astype(int)
    # Remove the selected pixel from the list
    L[index,:] = L[neg_pos,:]
    neg_pos = neg_pos - 1


# Pixels marked with 2 form the segmented región
J = J > 1
J = (J*255).astype(np.uint8)
cv_imshow(J)
```

References

1. Gonzalez, R. C., & Woods, R. E. (2018). Digital image processing (4th ed.). Pearson.
2. Otsu, N. (1979). A threshold selection method from gray-level histograms. *IEEE Transactions on Systems, Man, and Cybernetics, 9*(1), 62–66.
3. Haralick, R. M., & Shapiro, L. G. (1985). Image segmentation techniques. *Computer Vision, Graphics, and Image Processing, 29*(1), 100–132.
4. Adams, R., & Bischof, L. (1994). Seeded region growing. *IEEE Transactions on Pattern Analysis and Machine Intelligence, 16*(6), 641–647.
5. Pal, N. R., & Pal, S. K. (1993). A review on image segmentation techniques. *Pattern Recognition, 26*(9), 1277–1294.
6. Russ, J. C. (2011). The image processing handbook (6th ed.). CRC Press.
7. Sonka, M., Hlavac, V., & Boyle, R. (2014). Image Processing, analysis, and machine vision (4th ed.). Cengage Learning.

Chapter 10
Morphological Operations

This chapter provides a comprehensive exploration of mathematical morphology and its application to digital image processing. It begins by highlighting the limitations of traditional filters such as the median filter, which lack control over the preservation of specific geometric structures. Morphological operations overcome this by employing a structuring element, a shape model that enables the controlled manipulation of image components according to their size and form. The chapter introduces the fundamental operations of erosion (contraction) and dilatation (expansion), explaining their effects on object boundaries and their dual relationship. It details how combining these basic operations yields more advanced transformations such as opening and closing, which are used for noise removal, boundary smoothing, and structural restoration, and discusses their key mathematical properties, including idempotence and duality. Additionally, the hit-or-miss transformation is presented as a pattern detection technique for identifying specific pixel configurations. The text extends these principles to grayscale morphology, redefining erosion and dilatation in terms of intensity extrema, and introduces the morphological gradient for edge enhancement and the alternating sequential filter for progressive noise reduction. Finally, the chapter provides detailed Python implementations using OpenCV, illustrating the practical use of morphological operations for tasks such as edge detection, denoising, and shape analysis. Through theoretical foundations, geometric interpretation, and computational examples, the chapter establishes morphology as a robust framework for structure-based image analysis.

In Chap. 5, the behavior of the median filter was examined, highlighting that its application can significantly alter the two-dimensional structures present in an image. Such alterations typically manifest as the smoothing or rounding of corners, the suppression of thin elements like lines, or even the removal of small and isolated artifacts (Fig. 10.1). These effects arise because the median filter processes pixel values based on the intensity distribution within a local neighborhood, which makes it inherently sensitive to the geometric characteristics of nearby structures. While this selectivity allows the filter to adapt its response to certain shapes, its operation lacks

© The Author(s), under exclusive license to Springer Nature Switzerland AG 2026
E. Cuevas et al., *Image Processing with Python*, Signals and Communication
Technology, https://doi.org/10.1007/978-3-032-13285-7_10

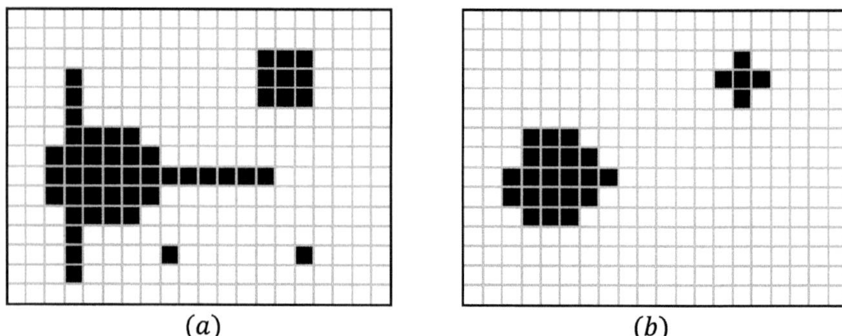

(a)   (b)

Fig. 10.1 Application of a 3×3 median filter on a binary image. **a** Original image and **b** filtered image using the median filter

precise control: it does not permit the user to explicitly target or preserve structures of a particular form or orientation.

To address this limitation, morphological filters are introduced in this chapter. Unlike the median filter, these techniques are grounded in mathematical morphology and operate through the definition of a structuring element, which serves as a model of the shape to be analyzed or modified within the image. By designing and selecting the appropriate structuring element, morphological filters provide a powerful and flexible framework to manipulate specific image features in a controlled manner, enabling tasks such as edge preservation, noise removal, and the enhancement or suppression of structures according to their geometry.

Morphological filters [1] were initially developed for use with binary images, where each pixel assumes only two possible values: 1 or 0, typically representing black and white. Such images are prevalent in numerous applications, particularly in document analysis and processing, where operations like character recognition, text segmentation, or noise elimination rely heavily on binary representations. In this binary context, morphological operations treat pixels with value 1 as belonging to the foreground or the structure of interest, while pixels with value zero are considered part of the background.

Although originally limited to binary data, the principles of morphological filtering have been extended to grayscale images, significantly broadening their applicability. In grayscale morphology, pixel intensities are no longer restricted to two discrete states but instead represent a continuous range of values. This adaptation enables morphological operators to manipulate more complex structures, allowing tasks such as texture analysis, feature extraction, edge detection, and noise reduction, while maintaining the ability to emphasize or suppress details based on shape.

Throughout this chapter, morphological filters will be examined in both contexts: their classical role in binary image processing and their extended functionality in grayscale domains, highlighting the flexibility and robustness of these techniques for different image analysis tasks.

10.1 Contraction and Growth of Structures

Considering the discussion above, the 3×3 median filter can also be applied to binary images. When used in this context, it tends to smooth and round the contours of large structures while simultaneously eliminating smaller objects or thin details. In practice, this means that the median filter suppresses structures whose size is below the window dimension used for filtering, in this case, smaller than a 3×3 neighborhood [2].

This observation raises a crucial question: how can we leverage the definition of a structure's size and shape to selectively and controllably influence specific image components? While the median filter produces interesting effects, such as removing small structures, it also has the disadvantage of unintentionally modifying or deforming larger ones. This lack of control limits its usefulness when structural preservation is critical.

To address this limitation, a more powerful approach emerges from the principles of mathematical morphology. Instead of relying solely on statistical measures like the median, morphology operates by systematically reducing and then expanding image structures using a predefined structuring element. The process can be summarized as follows:

1. Contraction (Erosion): All image structures are reduced. In this stage, small structures are completely removed, while larger ones shrink in size.
2. Elimination of Small Structures: As a result of the contraction, only objects larger than the structuring element survive, ensuring selective filtering.
3. Expansion (Dilatation): The reduced structures are regrown, allowing larger structures to recover their original size and shape, while the previously eliminated small structures do not reappear.

This sequence of contraction followed by expansion forms the basis of morphological opening, a process that preserves significant structures while eliminating irrelevant details, see Fig. 10.2. Unlike the median filter, morphological operators provide a controllable mechanism, since their behavior depends directly on the size and geometry of the chosen structuring element.

From the analysis of Fig. 10.2, it becomes evident that the removal of small structures in an image can be achieved through the definition of only two fundamental morphological operations. The first is Contraction (Erosion), which progressively removes pixels located at the boundary of an object, specifically those in direct contact with background pixels. As illustrated in Fig. 10.3, this process results in a reduction of the object's size, eventually leading to the disappearance of structures that are smaller than the chosen structuring element.

Conversely, the second operation, referred to as Growth (Dilatation), performs the opposite effect. In this case, a new layer of pixels is added to the boundary of the object, expanding it outward into the background region. As shown in Fig. 10.4, dilatation causes the object to increase in size and can also reconnect neighboring structures that are separated by small gaps.

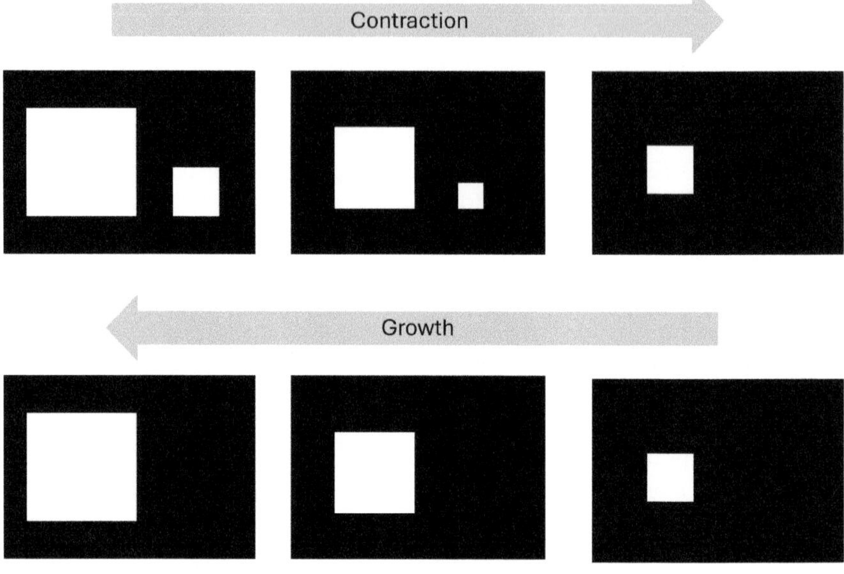

Fig. 10.2 Removal of small artifacts by contracting and growing structures in an image

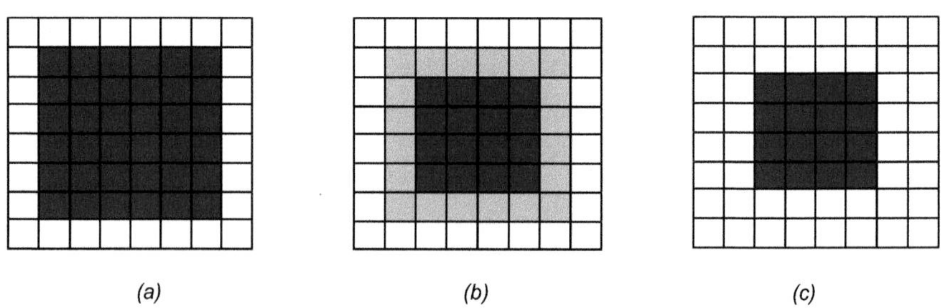

Fig. 10.3 Contraction of an image region. **a** Binary image, **b** boundary pixels into the surrounding background, and **c** structure boundary pixel removal

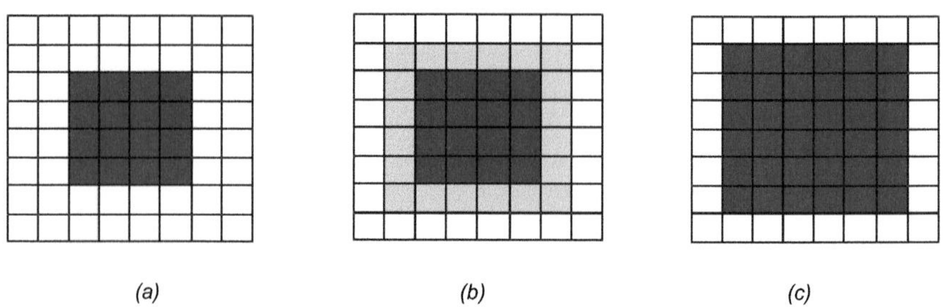

Fig. 10.4 Growth of an image region. **a** Binary image, **b** boundary pixels into the surrounding background, and **c** increase of the structure boundary

Together, erosion and dilatation constitute the core of mathematical morphology, and by combining them in sequence, more complex transformations such as opening and closing are derived. These combinations allow for precise and controllable manipulation of image structures: while erosion eliminates noise and small elements, dilatation restores or emphasizes the shapes of larger objects. This duality makes them powerful tools for structure-preserving image processing tasks.

10.1.1 Types of Neighborhoods

When performing morphological operations such as contraction (erosion) or growth (dilatation), it is essential to establish how the neighborhood relationship between pixels is defined, since this determines how structures within the image are modified. In digital image processing, two commonly used neighborhood models are considered:

- 4-Neighborhood (N4): In this scheme, a pixel is said to be in the neighborhood of another if it is directly adjacent either horizontally or vertically (above, below, left, or right). This definition results in each pixel having a maximum of four neighbors, forming a cross-shaped connectivity pattern, see Fig. 10.5a.
- 8-Neighborhood (N8): This neighborhood extends the N4 definition by including the diagonal neighbors as well. Thus, in addition to the four horizontal and vertical neighbors, the four diagonal pixels are also considered connected, yielding a maximum of eight neighbors arranged in a square pattern around the central pixel, see Fig. 10.5b.

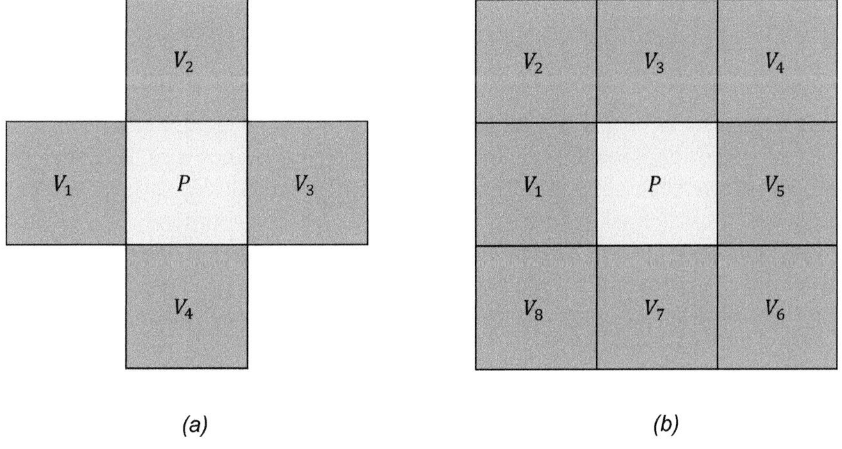

Fig. 10.5 Types of neighborhoods of the pixel P. **a** 4-neighborhood, and **b** 8-neighborhood

The choice between N4 and N8 connectivity has a direct impact on morphological operations:

- Using N4 may preserve thinner structures and enforce stricter connectivity rules.
- Using N8 allows for more flexible connectivity, facilitating smoother modifications of objects and potentially faster elimination of small, isolated components.

10.2 Fundamentals of Morphological Operations

The processes of reduction and growth form the basis of morphological filtering, as they correspond directly to the fundamental operations of erosion and dilatation in mathematical morphology. These operations are conceptually linked to the qualitative ideas discussed previously: erosion is associated with the loss of pixels along the boundaries of an object, effectively reducing its area or volume, while dilatation corresponds to the addition of pixels around the boundary, thereby expanding the object. In practical terms, erosion can be interpreted as the removal of a layer of boundary pixels from a structure, which leads to the contraction or thinning of objects, while dilatation adds a layer of pixels to the boundary, causing objects to grow or fill gaps.

This duality between erosion and dilatation creates a powerful analogy with natural physical processes: erosion models the gradual wearing down or shrinking of materials, while dilatation simulates expansion or growth. Within digital image processing, these operations allow for the controlled modification of object shapes, making them fundamental for applications such as noise removal, boundary smoothing, and structure extraction.

10.2.1 The Reference Structure

In a similar way to how conventional filters rely on a coefficient matrix to operate, morphological filters require the specification of a matrix known as the structuring element or reference structure. This matrix plays a central role in morphological operations, as it defines the shape and extent of the neighborhood used to probe and transform the image. Like binary images, the reference structure is composed exclusively of elements with values 0 and 1 (see Expression 10.1), where the ones determine the pixels that actively participate in the operation, while the zeros correspond to inactive positions.

$$H(i,j) \in \{0, 1\}. \tag{10.1}$$

Unlike the coefficient matrix of linear filters, which is typically centered at its origin, the structuring element incorporates the concept of a reference point

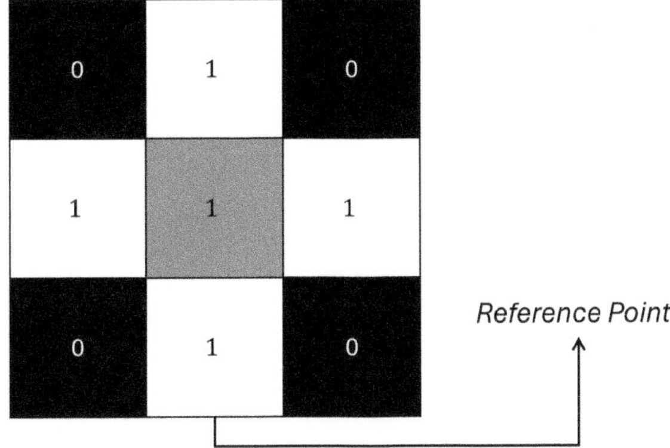

Fig. 10.6 Structuring element with cross-shape

(see Fig. 10.6). This point determines how the structuring element is aligned with the image during processing and does not necessarily have to be located at the center of the matrix. This flexibility provides a greater degree of freedom for designing structuring elements that can emphasize or suppress specific features in the image, such as corners, edges, or thin structures. Consequently, the choice of the structuring element's size, shape, and reference point directly influences the outcome of morphological operations, making it a crucial design parameter in image analysis tasks.

10.2.2 Set of Points

For the formal mathematical definition of morphological operations, it is often more convenient to represent images as sets of points identified by their two-dimensional coordinates, rather than as pixel matrices. In the case of a binary image $I(x, y)$, the image can be expressed as a set of points P_I, where each element corresponds to the coordinates of a pixel whose value is equal to 1. This formulation, introduced in Eq. 10.2, provides a compact and flexible representation of binary images.

$$P_I = \{(x, y \,|\, I(x, y) = 1\}. \tag{10.2}$$

As illustrated in Fig. 10.7, the same set-based notation can also be applied to the structuring element, allowing both the image and the reference structure to be treated uniformly within the same mathematical framework.

Using this representation, common binary image operations can be expressed in a straightforward way. For instance, the inversion of a binary image, which exchanges pixels with value 1 for 0 and vice versa, is equivalent to computing the complement of the set of points, as defined in Eq. 10.3, where $-I$ denotes the inversion of I.

$$P_{-I} = \overline{P_I}. \tag{10.3}$$

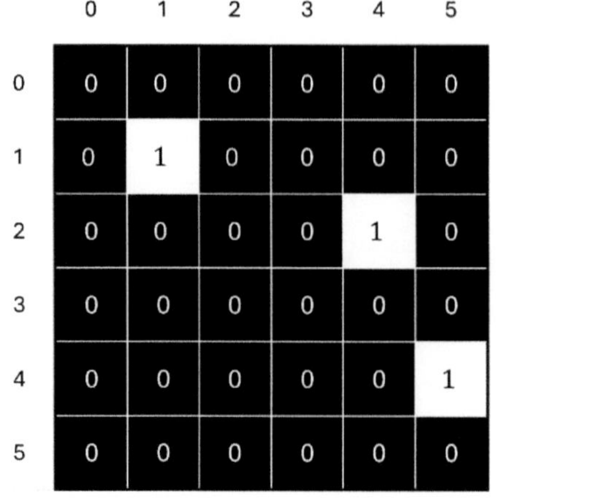
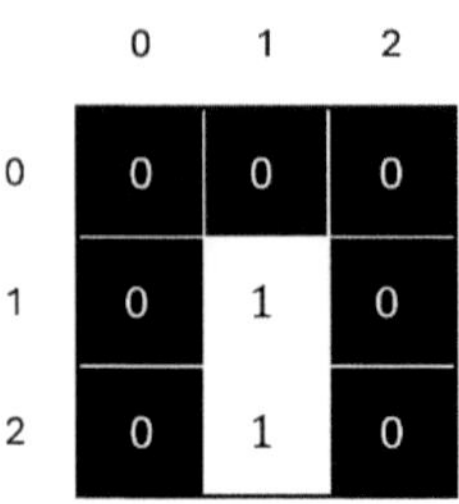

$$P_I = \{(1,1),(2,4),(4,5)\} \qquad\qquad P_H = \{(1,1),(2,1)\}$$

Fig. 10.7 Set of points P_I and P_H, where the set P_I corresponds to the pixels in the image and the set P_H corresponds to the elements of the reference structure

Likewise, if two binary images I_1 and I_2 are combined element by element through the logical OR operator, their union can be represented in set notation as in Eq. 10.4.

$$P_{I_1 \vee I_2} = P_{I_1} \cup P_{I_2}. \tag{10.4}$$

This duality between pixel-based logical operations and set-theoretic operations makes the set representation a powerful tool for describing morphological transformations. Depending on the context, the notation will be used interchangeably to simplify the explanation of algorithms and their implementation. For example, the union operation $I_1 \cup I_2$ is equivalent to $P_{I_1} \cup P_{I_2}$, and the complement $\bar{I}$ corresponds to $\overline{P_I}$. This flexibility ensures that the theoretical framework remains both rigorous and intuitive, while still being directly applicable to digital image processing tasks.

10.2.3 Dilatation

Dilatation is one of the fundamental morphological operations and can be understood as the formalization of the intuitive concept of growth, where an image structure is expanded by adding a layer of pixels around its boundary [3]. The extent and direction of this growth are governed by a geometric figure called the structuring element (or reference structure). In terms of set theory notation, dilatation is formally defined in Eq. 10.5.

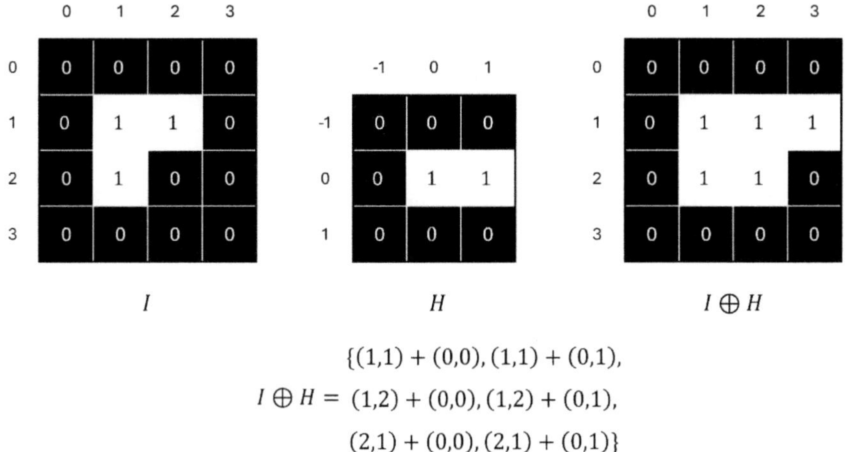

$$\begin{array}{rl}
& \{(1,1) + (0,0), (1,1) + (0,1), \\
I \oplus H = & (1,2) + (0,0), (1,2) + (0,1), \\
& (2,1) + (0,0), (2,1) + (0,1)\}
\end{array}$$

Fig. 10.8 Illustration of binary image I dilatation using the reference structure H

$$I \oplus H = \{(x', y') = (x + i, y + j) | (x', y') \in P_I, (i, j) \in P_H\}. \tag{10.5}$$

According to this definition, the dilatation of a binary image I by a structuring element H generates a new set of points that includes all possible coordinate combinations between the elements of the image set P_I and the structuring element set P_H. In other words, dilatation can be interpreted as the operation in which every pixel belonging to the image foreground P_I is replaced or "expanded" by the translated shape of the structuring element.

From a geometric perspective, this means that each foreground pixel of the original image acts as a reference point around which the structuring element is superimposed, effectively growing the object by the spatial configuration of H. As a result, small gaps within structures may be closed, thin regions may become thicker, and neighboring objects may merge if the dilatation exceeds their separation distance. This interpretation is visually exemplified in Fig. 10.8, where the dilatation process clearly illustrates how the structuring element determines the way in which the image grows.

10.2.4 Erosion

Erosion is considered the quasi-inverse operation of dilatation and represents the morphological counterpart associated with the reduction of image structures [4]. In point set notation, erosion is formally defined in Eq. 10.6. According to this definition, for a pixel (x', y') in the image, the result of the erosion is determined by verifying whether all possible shifted coordinates $(x' + i, y' + j)$, corresponding to the structuring element, belong to the original image set I. Only in that case does the pixel (x', y') remain part of the eroded image.

$$I \ominus H = \{(x', y') \mid (x' + i, y' + j) \in P_I(i,j), \forall (i,j) \in P_H\}. \tag{10.6}$$

Intuitively, this process can be interpreted as follows: a pixel is preserved as foreground (value 1) in the eroded image if, when the structuring element is centered on that pixel, the shape of the structuring element is fully contained within the original image region. If any part of the structuring element extends outside the object, the pixel is removed.

From a geometric standpoint, erosion progressively shrinks objects, reducing their area and eliminating narrow protrusions or fine details. This makes it particularly effective for removing small noise artifacts or separating objects that are weakly connected. However, excessive erosion can also fragment larger structures or even eliminate them entirely if the structuring element is too large.

Figure 10.9 illustrates this principle, showing an example where the structuring element H is superimposed on the binary image I. Specifically, the figure highlights how, when the structuring element is centered on pixel $(1,1)$, it perfectly matches the image content, resulting in that pixel being preserved (value 1) in the eroded image.

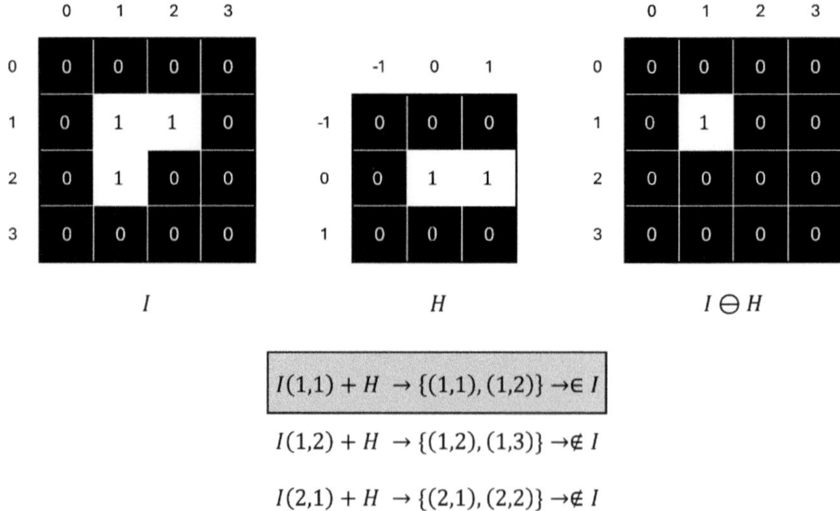

$$I(1,1) + H \rightarrow \{(1,1), (1,2)\} \rightarrow \in I$$

$$I(1,2) + H \rightarrow \{(1,2), (1,3)\} \rightarrow \notin I$$

$$I(2,1) + H \rightarrow \{(2,1), (2,2)\} \rightarrow \notin I$$

Fig. 10.9 Erosion example applied to the binary image I, using the reference structure H

10.2.5 *Dilatation and Erosion Properties*

Although dilatation and erosion are closely related operations, they cannot be considered strict mathematical inverses [5]. In practice, an image that has been eroded cannot be perfectly reconstructed by successive dilatations, and the reverse situation also holds. Nevertheless, these operators exhibit a strong duality: performing a dilatation on the foreground pixels (those with value 1) is equivalent to performing an erosion on the background pixels (those with value 0). This relationship is formalized in Eq. 10.7.

$$\bar{I} \ominus H = \overline{I \oplus H}. \tag{10.7}$$

From an algebraic standpoint, dilatation possesses several important properties. First, dilatation is commutative, meaning that the order of the operands (the image or the structuring element) can be exchanged without altering the result, as expressed in Eq. 10.8. This property makes dilatation analogous to convolution in linear systems, where the interchange of kernel and input signal is also valid.

$$I \oplus H = H \oplus I. \tag{10.8}$$

Furthermore, dilatation is associative, which implies that when applying multiple dilatations with different structuring elements, the grouping of operations is irrelevant. This is expressed in Eq. 10.9.

$$(I_1 \oplus I_2) \oplus I_3 = I_1 \oplus (I_2 \oplus I_3). \tag{10.9}$$

The associativity property has practical implications for computational efficiency. Specifically, a large structuring element H can be decomposed into smaller structuring elements H_1 and H_2, so that the dilatation with H can be achieved by successive dilatations with H_1 and H_2. This decomposition reduces the number of operations and increases processing speed. Mathematically, this is expressed in Eq. 10.10.

$$I \oplus H = ((I \oplus H_1) \oplus H_2). \tag{10.10}$$

Erosion, on the other hand, does not share all the algebraic properties of dilatation. In particular, erosion is not commutative, which means that the result of applying erosion depends on the order in which the image and the structuring element are considered. This distinction is formalized in Eq. 10.11.

$$I \ominus H \neq H \ominus I. \tag{10.11}$$

10.2.6   Morphological Filter Design

Morphological filters are formally characterized by two elements: the operation to be performed (either erosion or dilatation) and the corresponding structuring element, also referred to as the reference structure. The size and shape of this reference structure are not arbitrary but depend directly on the application requirements. In practice, a set of standard structuring elements is widely used, as illustrated in Fig. 10.10.

For instance, when a disk-shaped structuring element of radius r is employed, its application through dilatation causes the objects in the image to expand by adding an external layer of r pixels. Conversely, when the same structuring element is used for erosion, it produces the opposite effect: the boundary of the objects contracts by removing a layer of r pixels. This effect is exemplified in Figs. 10.11 and 10.12, where the original image (Fig. 10.11a) undergoes dilatation and erosion using the disk structuring element of Fig. 10.11b, with varying values of r.

Similarly, Fig. 10.13 illustrates how structuring elements of different shapes give rise to distinct morphological transformations.

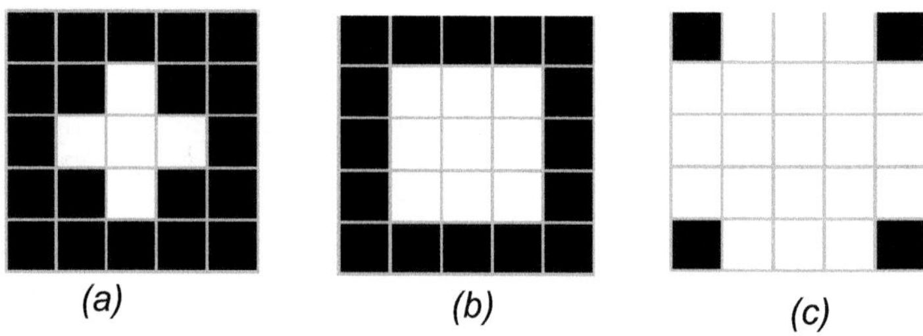

(a)   (b)   (c)

Fig. 10.10 Different reference structures: **a** cross structure, **b** rectangle structure, and **c** small disk structure

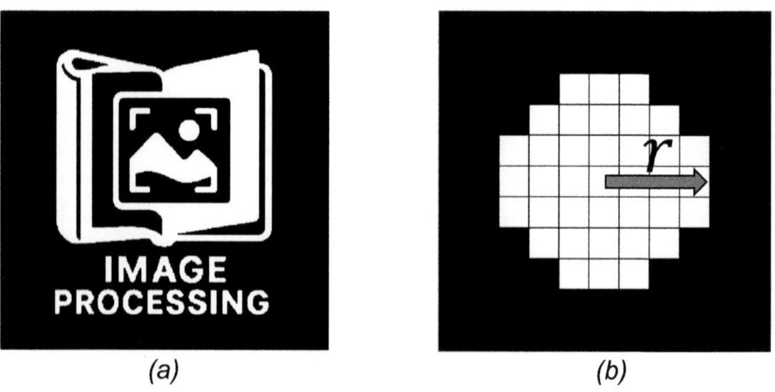

(a)   (b)

Fig. 10.11 **a** Binary image and **b** disk reference structure

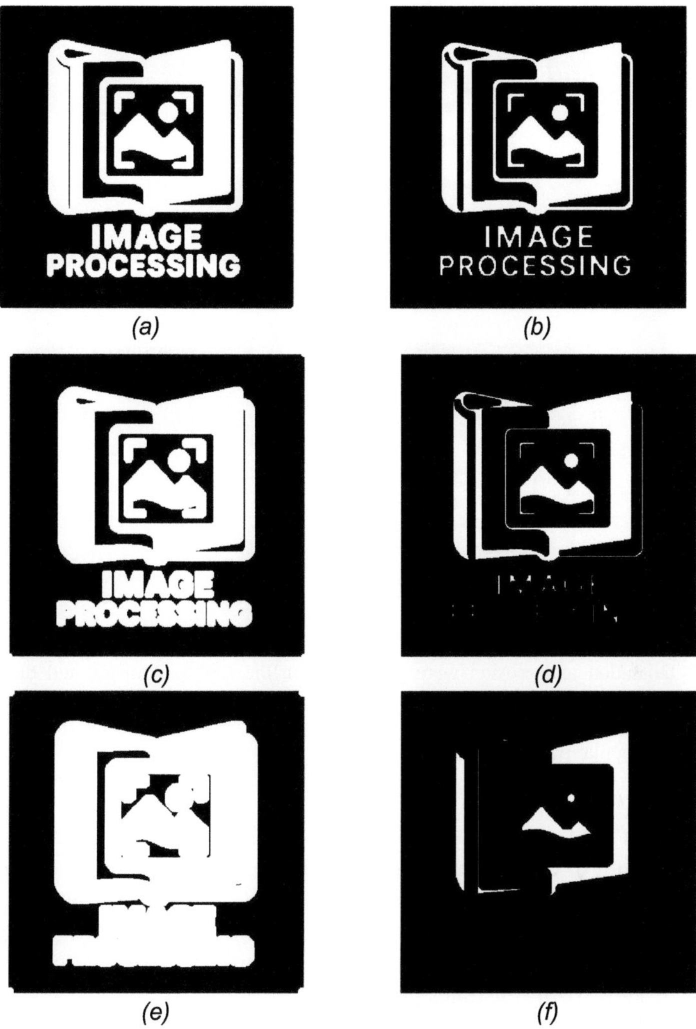

Fig. 10.12 Example of dilatation and erosion operations considering different values of r for the size of the disk reference structure. **a** Dilatation and **b** erosion considering $r = 2$; **c** dilatation and **d** erosion considering $r = 5$; and **e** dilatation and **f** erosion considering $r = 10$

Unlike spatial filtering techniques, it is generally not possible to construct isotropic two-dimensional morphological filters by combining one-dimensional structuring elements H_x and H_y. This limitation arises because the morphological operation between two reference structures tends to generate a square-shaped structuring element, which inherently lacks rotational invariance.

When morphological filters must be applied with large structuring elements, a practical approach to reduce computational complexity is to perform iterative

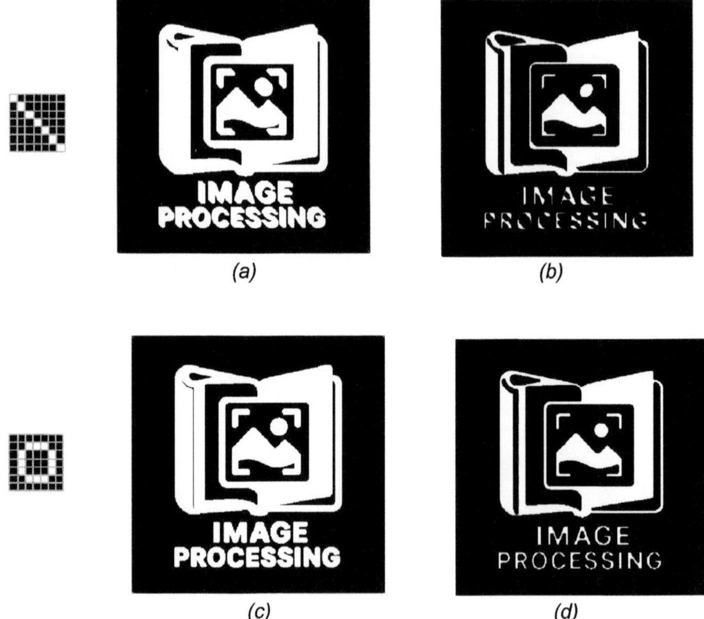

Fig. 10.13 Dilatation and erosion were performed on the example image, and the reference structure was used, as shown on the left side. **a** Dilatation and **b** erosion are used as reference structure values of 1 in the main diagonal, and **c** dilatation and **d** erosion are used as reference structures for a circular pattern

applications of the same structuring element with smaller dimensions. In this scheme, the result of one operation is reused as the input for the next, progressively accumulating the effect of a larger structuring element. This iterative method yields results that closely approximate those obtained using a single, high-dimensional structuring element, while significantly improving computational efficiency. Figure 10.14 illustrates this principle.

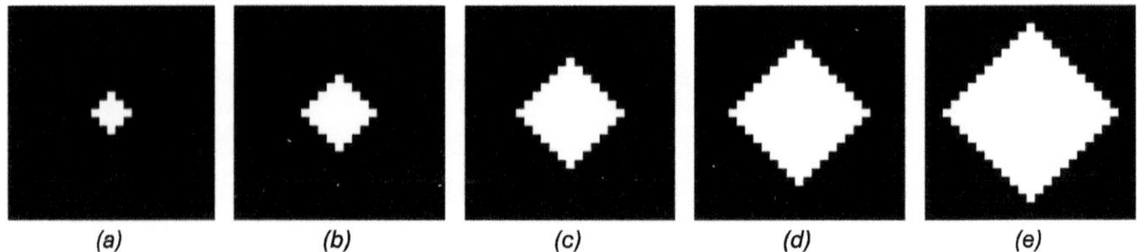

Fig. 10.14 Formation of large filters by successively applying a filter H to the image. Example for dilatation: **a** Application of 1, **b** 2, **c** 3, **d** 4, and **e** 5 filters successively

10.3 Edge Detection in Binary Images

One of the most common applications of morphological operations is the extraction of object boundaries in binary images. The detection procedure, summarized in Algorithm 10.1, relies on the sequential application of erosion, inversion, and logical operations.

The process begins by applying the erosion operation (Step 1) to the original binary image, using as a structuring element one of the reference structures shown in Fig. 10.10, which encodes neighborhood relationships. The effect of erosion at this stage is to remove the outermost layer of pixels from each object in the image, effectively shrinking them and discarding their boundaries.

Once the eroded version of the image is obtained, the next step (Step 2) is to compute its complement (inverse). In this transformed image, the interior of the objects is represented by pixels with a value of 0, while the background and the removed boundary pixels are represented by 1.

Finally, an AND logical operation (Step 3) is performed between the original image and the inverted eroded version. Since the only overlapping pixels between these two images correspond to the object boundaries, the result of this operation is a binary image where the edges are clearly extracted and highlighted.

This process, illustrated in Fig. 10.15, demonstrates how simple morphological transformations, when properly combined, provide an efficient method for edge detection in binary imagery.

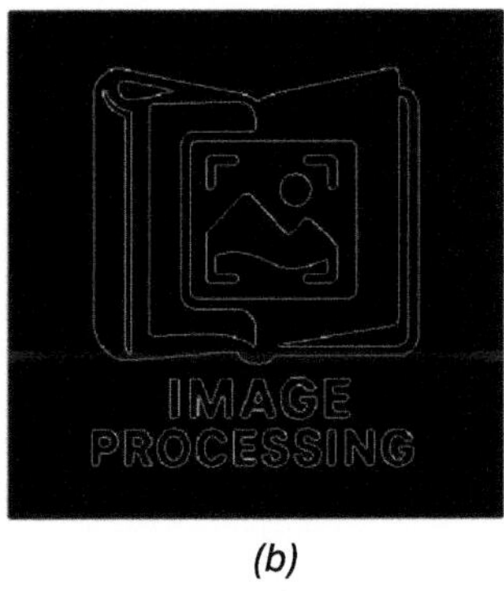

(a)   (b)

Fig. 10.15 Erosion for edge detection using algorithm 10.1. **a** Original image and **b** edge detection result

Algorithm 10.1 Algorithm for edge detection using morphological operations

	Edge detection by erosion ($I_b(x, y)$) where $I_b(x, y)$ is a binary image
1:	$I_b(x, y))$ is eroded by using one of the following structuring elements to obtain I' $I'(x, y) = I \ominus H$
2:	The eroded image is inverted $I_{inv}(x, y) = \overline{I'(x, y)}$
3:	The AND operation is applied between $I_{inv}(x, y)$ and $I_b(x, y)$ $I_{edge}(x, y) = I_{inv}(x, y) \wedge I_b(x, y)$

10.4 Combination of Morphological Operations

In practical image processing applications, the fundamental morphological operations of dilatation and erosion are rarely used in isolation. Instead, they are typically applied in combinations that enhance their usefulness in tasks such as noise reduction, shape analysis, and object detection. Among these, three operations are especially relevant: opening, closing, and the hit-or-miss transformation.

Because of the quasi-dual relationship between dilatation and erosion, these two operations can be composed in different orders to yield transformations with distinct properties. Two of these combinations are of particular importance in mathematical morphology and have been assigned specific names and symbols: opening and closing.

Together, opening and closing form the basis for a wide range of morphological filtering techniques. Their complementary effects make them powerful tools for preprocessing and structural analysis of binary and grayscale images.

10.4.1 Opening

Opening is a morphological operation formally defined as the composition of an erosion followed by a dilatation, both performed with the same structuring element (or reference structure). Its mathematical formulation is expressed in Eq. 10.12.

$$I \circ H = (I \ominus H) \oplus H. \tag{10.12}$$

Conceptually, the process can be interpreted in two stages:

- Erosion Stage: The image is first eroded, which eliminates all foreground pixels (value 1) corresponding to objects or structures that are smaller than the structuring element. In this step, thin protrusions, noise points, and small irregularities are removed, leaving only the portions of objects that can fully contain the reference structure.
- Dilatation Stage: After erosion, dilatation is applied with the same reference structure. This operation restores the size of the surviving objects, smoothing their contours and re-expanding them to approximately their original dimensions prior to erosion.

The combined effect of opening is to smooth object boundaries, eliminate noise components smaller than the structuring element, and preserve the main shapes of larger objects. For this reason, opening is commonly used as a preprocessing step in image analysis tasks, such as segmentation or shape recognition.

Figure 10.16 illustrates this effect using a disk-shaped structuring element with different radio r. As shown in Fig. 10.16b, increasing the value of r strengthens the filtering action, progressively removing finer details and retaining only structures large enough to accommodate the chosen structuring element.

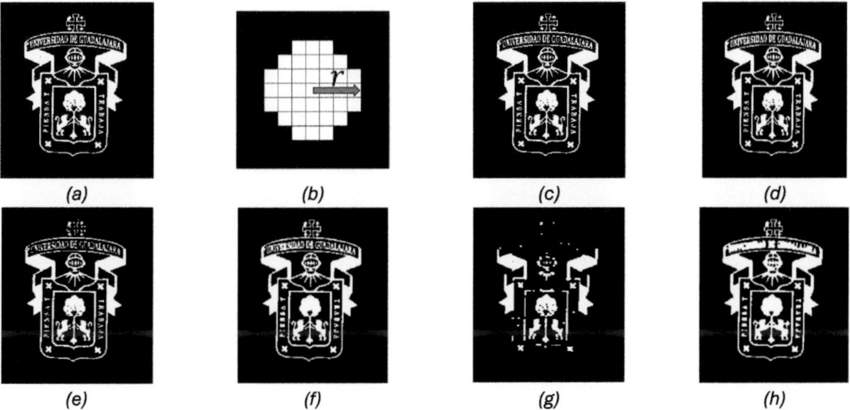

Fig. 10.16 Examples of opening and closing considering different reference structures' sizes, **a** binary image, **b** reference structure, **c** opening, and **d** closing with $r = 1$, **e** opening and **f** closing with $r = 3$, **g** opening and **h** closing with $r = 5$

10.4.2 Closing

Closing is the morphological operation defined as the composition of a dilatation followed by an erosion, both using the same structuring element. Its formal definition is expressed in Eq. 10.13.

$$I \cdot H = (I \oplus H) \ominus H. \tag{10.13}$$

Conceptually, the process can be described in two steps:

- Dilatation Stage: The image is first dilated, which causes the expansion of object boundaries. As a result, small gaps, narrow breaks, or holes within the objects that are smaller than the structuring element are closed.
- Erosion Stage: A subsequent erosion is applied with the same structuring element. This step restores the size of the objects to approximately their original dimensions, while preserving the holes that were already filled during dilatation.

The combined effect of closing is to smooth the contours of objects, fill small holes or gaps, and connect nearby components separated by thin spaces. Thus, closing is especially useful in applications where the goal is to reinforce the connectivity of objects and eliminate small background regions inside them.

Figure 10.16 illustrates the result of the closing operation when using a disk-shaped structuring element of radius r. As observed in Fig. 10.16b, increasing the radius r intensifies the effect, progressively merging closer structures and filling larger gaps.

10.4.3 Properties of Opening-and-Closing Operations

Both opening and closing present fundamental mathematical properties that make them highly useful in morphological image processing. These properties are Idempotence and duality.

In the Idempotence property, once an opening (or closing) operation has been applied to an image, performing the same operation repeatedly will not further change the result. This property is defined in Eq. 10.14, where o denotes opening and · denotes closing. This property guarantees stability because applying the same operation more than once is unnecessary, since the first application already provides the final result.

$$(I \circ H) \circ H = I \circ H$$
$$(I \cdot H) \cdot H = I \cdot H. \tag{10.14}$$

In the duality property, opening and closing are dual operations with respect to image complementation. Specifically, the opening of the foreground (pixels with

value 1) is equivalent to the closing of the background (pixels with value 0), and vice versa. This property is expressed in Eq. 10.15, where.
$\overline{(I \circ H)}$ denotes the complement of the image I.

$$I \circ H = \overline{(I \cdot H)}$$
$$I \cdot H = \overline{(I \circ H)}. \tag{10.15}$$

These properties confirm that opening and closing are not arbitrary combinations of erosion and dilatation, but well-structured operations with predictable, stable, and complementary behaviors.

10.4.4 The "Hit-or-Miss" Transformation

The hit-or-miss transform is a morphological operation particularly useful for detecting specific pixel configurations within a binary image. It is denoted as $I \otimes H$. Unlike opening and closing, which rely on a single reference structure, the hit-or-miss transform employs a pair of reference structures $H = (H_1, H_2)$, where H_1 defines the foreground pattern that must match (hit), and H_2 defines the background pattern that must not match (miss).

The transformation is defined in Eq. 10.16 and operates as follows:

1. Erosion ($I \ominus H_1$): Detects all pixel positions where the image matches the desired foreground structure (hit).
2. Erosion ($I^c \ominus H_2$): Detects all positions where the complementary background structure is absent (miss).
3. Intersection $\cap$: The final result is obtained by applying the logical AND between both eroded images, ensuring that the configuration occurs exactly as specified.

$$I \otimes H = (I \ominus H_1) \cap (I^c \ominus H_2). \tag{10.16}$$

Figures 10.17 and 10.18 illustrate the application of this transformation to detect a cross-shaped pixel configuration within a binary image.
$I \otimes H = (I \ominus H_1) \cap (I^c \ominus H_2)$ on Fig. 10.17c
The name "hit-or-miss" arises from its dual nature:

- A hit occurs when the pixels of the image coincide with H_1.
- A miss occurs when the pixels coincide with H_2.

This property makes the hit-or-miss transformation a fundamental tool for pattern detection in morphological image processing.

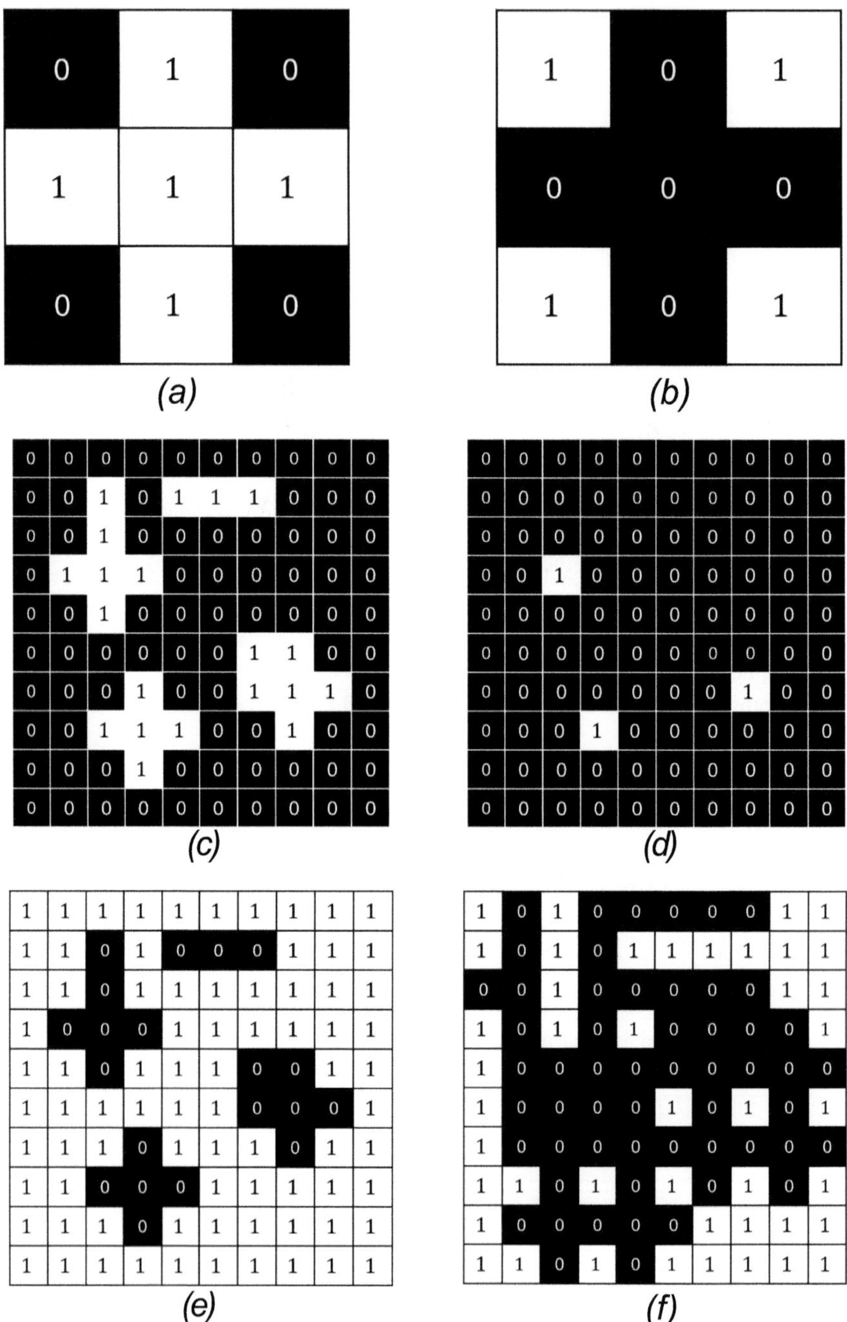

Fig. 10.17 Hit-or-miss transformation process considering **a** the reference structure H_1, **b** the reference structure H_2, **c** the binary image, **d** the outcome of $I \ominus H_1$, **e** the complement of the binary image I^c, and **f** the outcome of $I^c \ominus H_2$

Fig. 10.18 Hit-or-miss transform outcome:

0	0	0	0	0	0	0	0	0	0
0	0	0	0	0	0	0	0	0	0
0	0	0	0	0	0	0	0	0	0
0	0	1	0	0	0	0	0	0	0
0	0	0	0	0	0	0	0	0	0
0	0	0	0	0	0	0	0	0	0
0	0	0	0	0	0	0	0	0	0
0	0	0	1	0	0	0	0	0	0
0	0	0	0	0	0	0	0	0	0
0	0	0	0	0	0	0	0	0	0

10.5 Morphological Operations for Grayscale Images

Morphological operations are not limited to binary images; they can also be extended to grayscale images. In fact, all the operations described earlier, except for the hit-or-miss transform, have a natural adaptation to this type of image.

In the case of color images, morphological processing is usually performed by treating each color channel (or plane) independently, as if it were a separate grayscale image.

Although the operations retain the same names and symbols as in the binary case, their definitions differ significantly when applied to intensity or color images, due to the nature of pixel values and their continuous range.

10.6 Reference Structure

The reference structure (or structuring element) is the first key distinction between morphological operations applied to binary images and those applied to grayscale or color images.

In the case of binary images, the reference structure is typically defined as a matrix of ones and zeros, where the ones describe the shape and dimensions of the structuring element that will interact with the image, while the zeros represent inactive positions.

For grayscale or intensity images, however, the reference structure takes on a more complex role. Instead of being limited to binary values, it is represented as a two-dimensional real-valued function, as defined in Expression 10.17. This allows the structuring element to incorporate positive, negative, or zero values, functioning similarly to the coefficient matrices used in spatial filtering.

$$H(i, j) \in \mathbb{R} \qquad (10.17)$$

A critical difference arises here: in morphological operations, unlike in linear spatial filters, the zero values of the structuring element are not neutral; they actively influence the outcome of the operation. This property makes the design of the reference structure particularly important, as it directly affects how features in the image are enhanced, suppressed, or preserved.

10.6.1 Dilatation and Erosion for Grayscale Images

In the context of grayscale morphological operations, dilatation ($\oplus$) and erosion ($\ominus$) are redefined to account for the intensity values of the image.

Grayscale dilatation is defined as the maximum value obtained from the sum of the pixel intensities in the neighborhood of the image and the corresponding coefficients of the reference structure. Mathematically, this is expressed in Eq. 10.18. Intuitively, dilatation brightens the regions where the structuring element overlaps with high-intensity pixels, thus expanding bright features in the image.

$$(I \oplus H)(x, y) = \max_{(i,j)\in H} \{I(x + i, y + j) + H(i, j)\}. \qquad (10.18)$$

On the other hand, grayscale erosion is defined as the minimum value resulting from the subtraction of the coefficients of the reference structure from the corresponding image region, as formulated in Eq. 10.19. This operation tends to darken the image and shrink bright structures, highlighting the darker components of the scene.

$$(I \ominus H)(x, y) = \min_{(i,j)\in H} \{I(x + i, y + j) + H(i, j)\}. \qquad (10.19)$$

Figures 10.19 and 10.20 illustrate the effects of grayscale dilatation and erosion, respectively.

Since these operations involve arithmetic computations, the resulting values may exceed the valid range of gray levels (typically 0–255 in 8-bit images). To address this, a limiting process called clamping is applied:

- Any negative result is set to 0.
- Any result greater than 255 is set to 255.

This ensures the processed image remains within the representable dynamic range.

Furthermore, dilatation and erosion can be combined to derive more complex transformations. A notable example is the morphological gradient, which is obtained

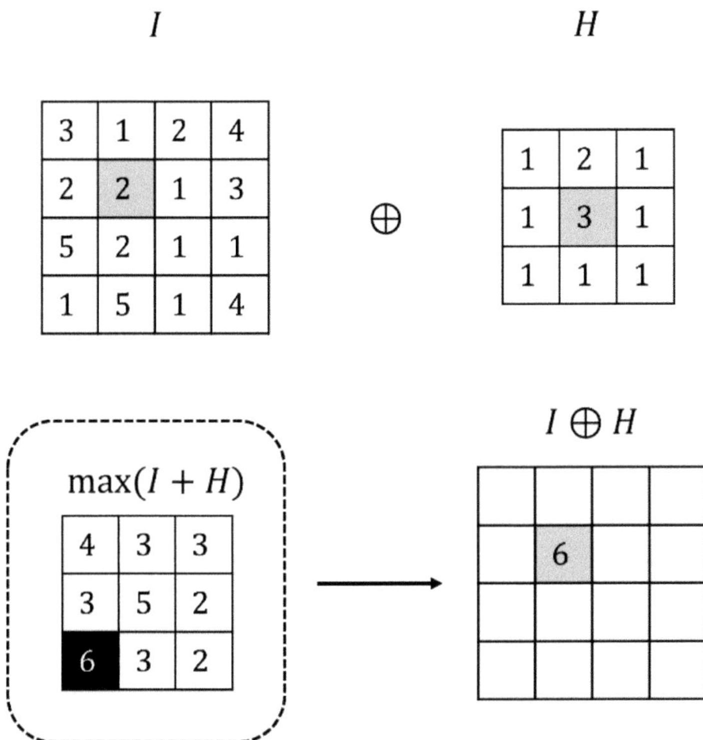

Fig. 10.19 Example of the process for grayscale image dilatation. The 3×3 reference structure H is centered on the image I. The sum operation $I + H$ is processed. From this operation, the maximum is chosen: $\max(I + H)$, which is considered the result of the dilatation process $I \oplus H$

by subtracting the eroded version of the image from its dilated version, as shown in Eq. 10.20.

$$\partial M = (I \oplus H) - (I \ominus H). \tag{10.20}$$

The morphological gradient highlights the edges and transitions of intensity in grayscale images, functioning as a powerful tool for feature detection. Figure 10.21 presents examples of dilatation, erosion, and the resulting morphological gradient.

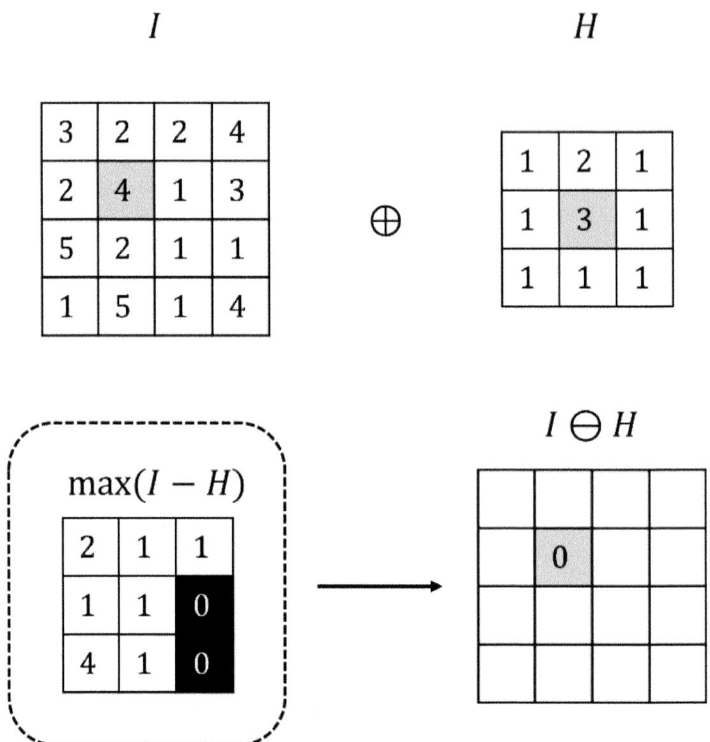

Fig. 10.20 Example of the process for grayscale image erosion. The 3×3 reference structure H is centered on the image I. The subtraction operation $I - H$ is processed. From this operation, the minimum is chosen: $\min(I - H)$, which is considered the result of the erosion process $I \ominus H$

10.6.2 Opening-and-Closing Operations for Grayscale Images

The opening-and-closing operations for grayscale images retain the same formal structure as their binary counterparts, but they act on intensity values rather than simple binary pixels. Both operations have clear geometric interpretations, which can be illustrated using an image profile (i.e., the intensity values along a line of the image).

Conceptually, opening can be visualized as sliding the reference structure along the bottom of the image profile. In Fig. 10.22, a horizontal structuring element (see Fig. 10.22b) is fitted under the intensity curve (see Fig. 10.22a). Small peaks that are narrower than the structuring element are removed, as shown in Fig. 10.22c. In general, the opening operation eliminates small bright details or spikes in the image, while larger intensity features remain largely unchanged.

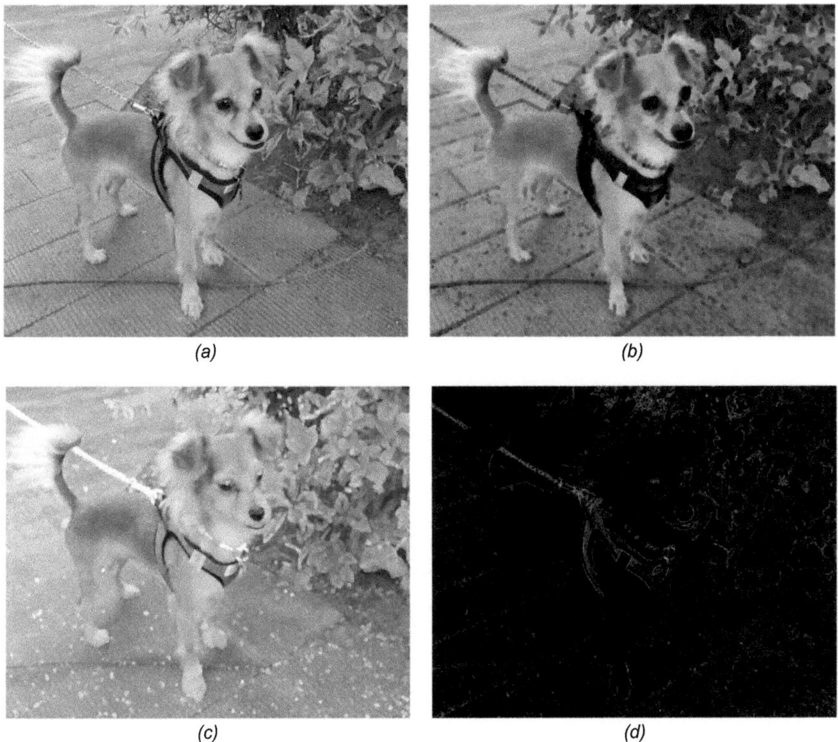

Fig. 10.21 Morphological gradient on grayscale images: **a** grayscale image, **b** eroded image, **c** dilated image, and **d** morphological gradient

Closing can be visualized as fitting the structuring element over the top of the image profile. Small valleys that are narrower than the structuring element are filled in, as shown in Fig. 10.22d, e. This operation removes small dark artifacts without significantly altering larger intensity features.

Because opening removes bright details and closing eliminates dark artifacts, combining the two operations is a powerful strategy for noise reduction in images.

By first applying an opening operation (to remove bright artifacts) and then a closing operation (to remove dark artifacts), this filter can simultaneously suppress salt-and-pepper noise. Figure 10.23 illustrates the effect of the opening-and-closing filter on a grayscale image corrupted with salt-and-pepper noise.

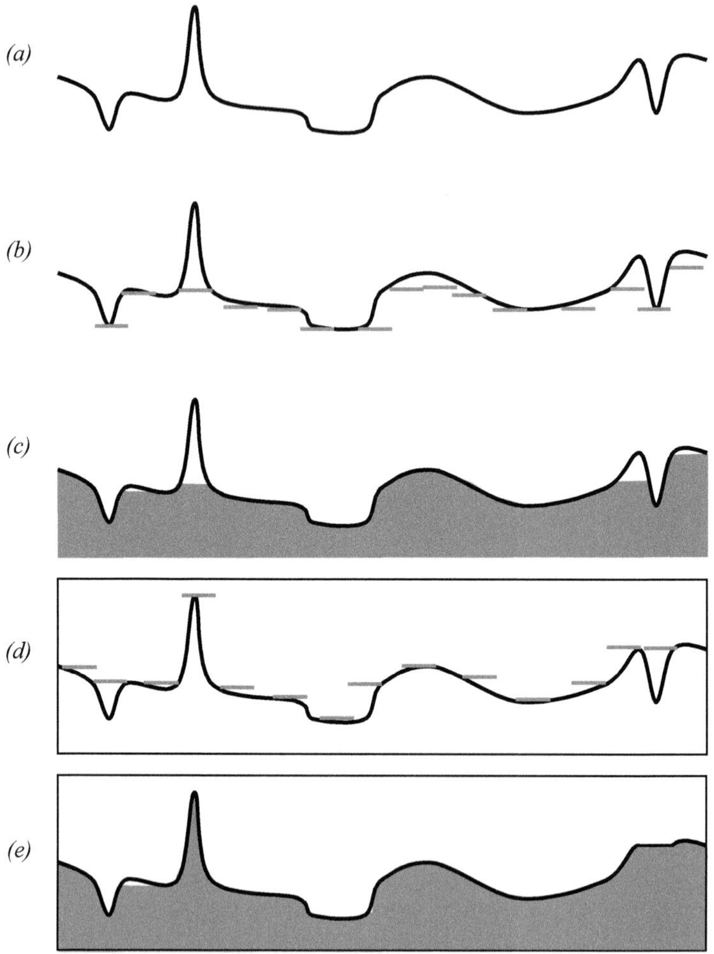

Fig. 10.22 Illustration of closing and opening for grayscale images: **a** profile of a grayscale image, **b** application of opening, **c** opening result, **d** application of closing, and **e** closing result

To enhance smoothing, the opening-and-closing filter can be applied repeatedly, with varying sizes of the reference structure in each iteration. This process is known as the alternating sequential filter, which produces smoother results compared to a single application of the opening-and-closing filter. Figure 10.24 shows the effect of the alternating sequential filter on a grayscale image, and Algorithm 10.2 details its implementation process.

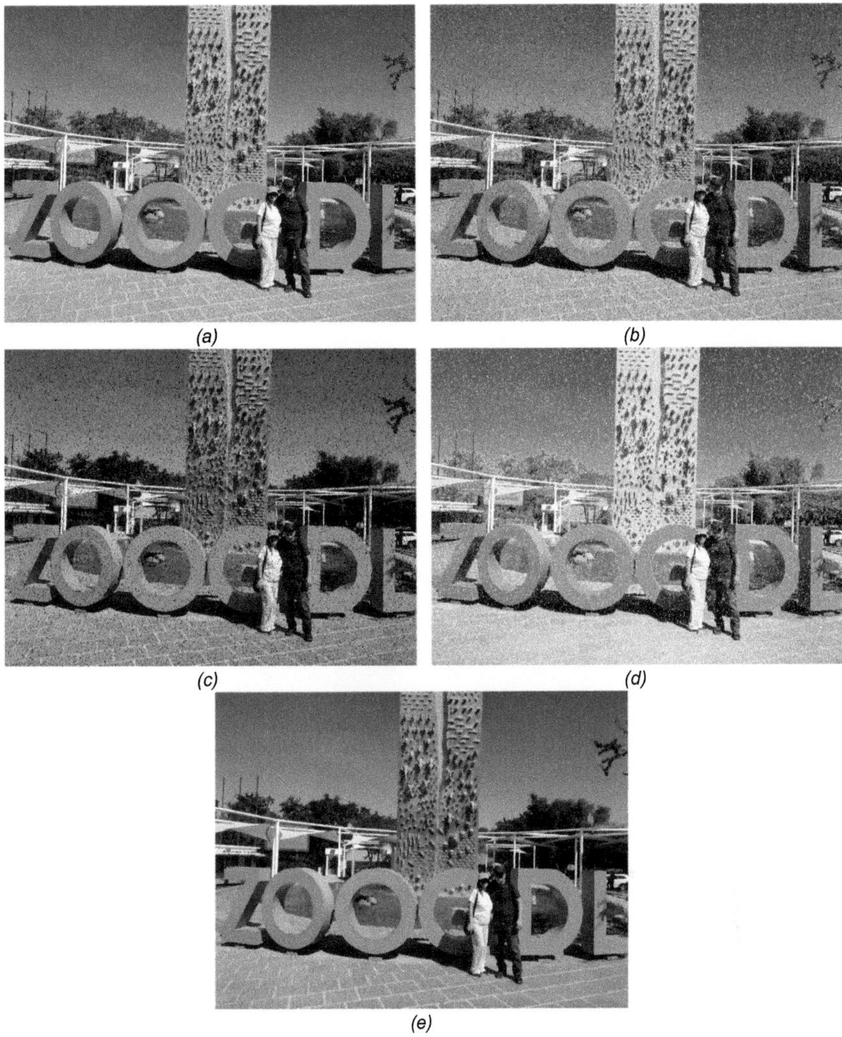

Fig. 10.23 Opening-and-closing operations for denoising grayscale images: **a** grayscale image, **b** grayscale image with salt-and-pepper noise, **c** noisy image after opening, **d** noisy image after closing, and **e** denoised image after opening-and-closing filter

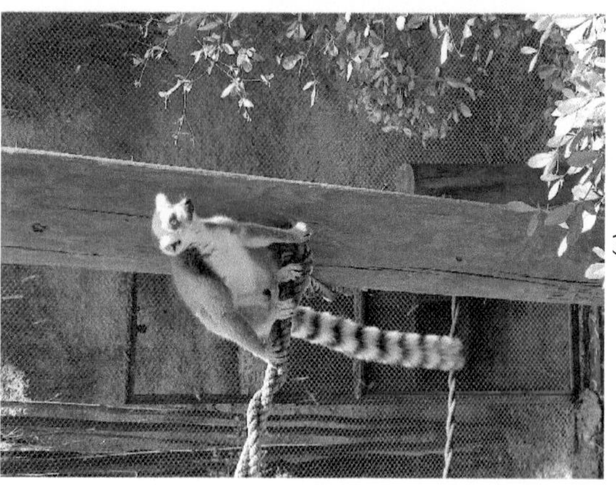

Fig. 10.24 Alternating sequential filter. **a** Grayscale image and **b** outcome using 3 repetitions considering r from 1 to 3

Algorithm 10.2 Sequential filter toggle implementation algorithm

	Alternating sequential filter $(I(x, y))$
	Where $I(x, y)$ is a grayscale image
1:	*forr* $= 1 : t$(for an appropriate value of r) 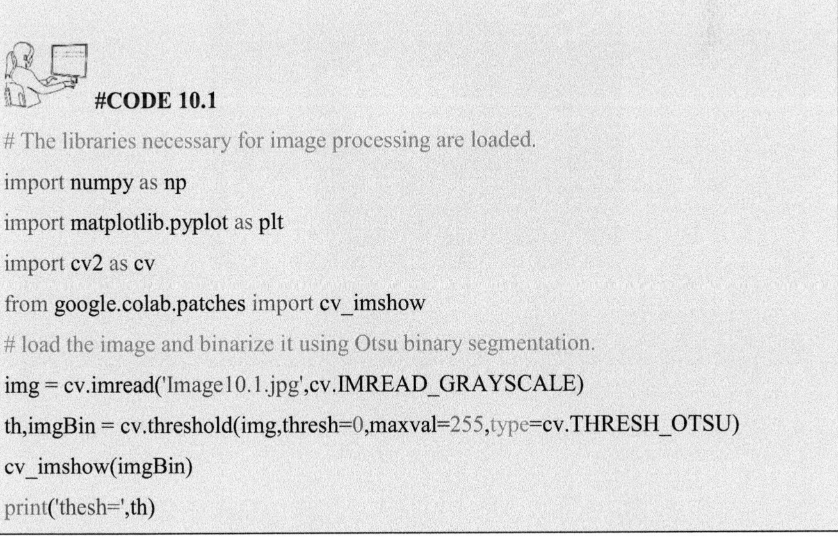
2:	Opening $I_1 = I \circ H$
3:	Closing $I_2 = I \cdot H$
4:	Assigns $I = I_2$
5:	*end* of *for*
6:	The final value of I expresses the result of the filter

10.7 Implementation of Morphological Operation in Python

> **Morphological Operation**
>
> **#CODE 10.1**
>
> ```python
> # The libraries necessary for image processing are loaded.
> import numpy as np
> import matplotlib.pyplot as plt
> import cv2 as cv
> from google.colab.patches import cv_imshow
> # load the image and binarize it using Otsu binary segmentation.
> img = cv.imread('Image10.1.jpg',cv.IMREAD_GRAYSCALE)
> th,imgBin = cv.threshold(img,thresh=0,maxval=255,type=cv.THRESH_OTSU)
> cv_imshow(imgBin)
> print('thesh=',th)
> ```

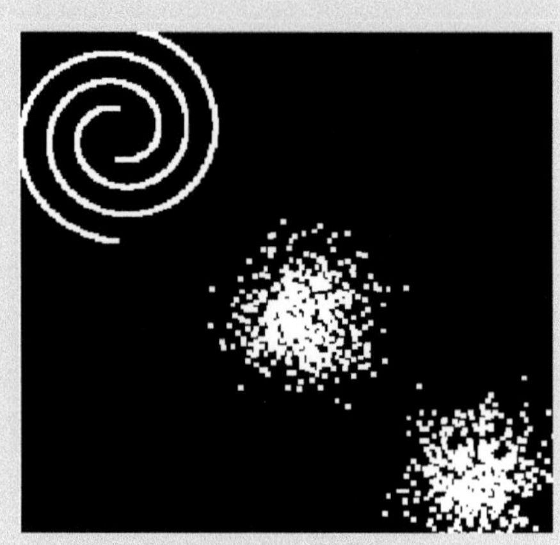

thesh= 39.0

Dilatation

#Ellipse 3x3

```python
imgDilated = cv.dilate(imgBin, cv.getStructuringElement(cv.MORPH_ELLIPSE, (3,3)))
cv_imshow(imgDilated)
```

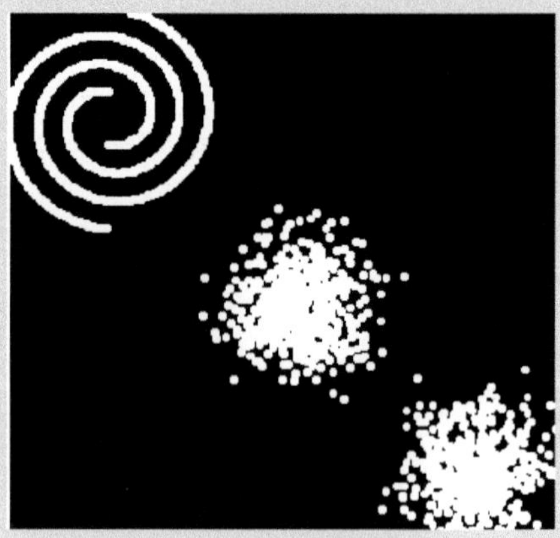

#Ellipse 9x9

```
imgDilated = cv.dilate(imgBin, cv.getStructuringElement(cv.MORPH_ELLIPSE, (9,9)))
cv_imshow(imgDilated)
```

```
#Ellipse 3x3, 5 iterations
imgDilated = cv.dilate(imgBin, cv.getStructuringElement(cv.MORPH_ELLIPSE, (3,3)),iterations=3)
cv_imshow(imgDilated)
```

```
#Rectangle 3x3
```

```
imgDilated = cv.dilate(imgBin, cv.getStructuringElement(cv.MORPH_RECT, (3,3)))
cv_imshow(imgDilated)
```

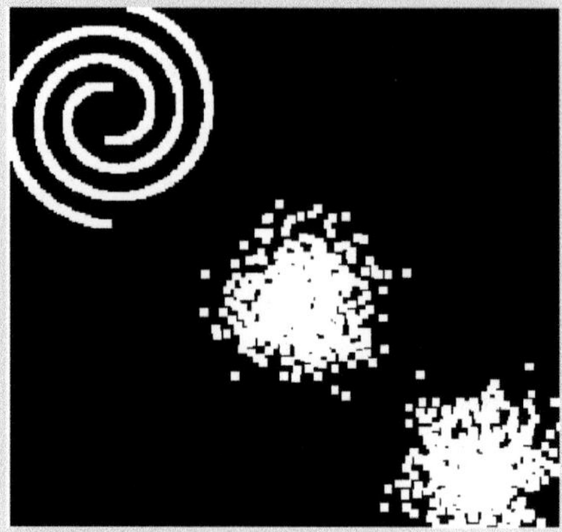

```
#Rectangle 9x9
imgDilated = cv.dilate(imgBin, cv.getStructuringElement(cv.MORPH_RECT, (9,9)))
cv_imshow(imgDilated)
```

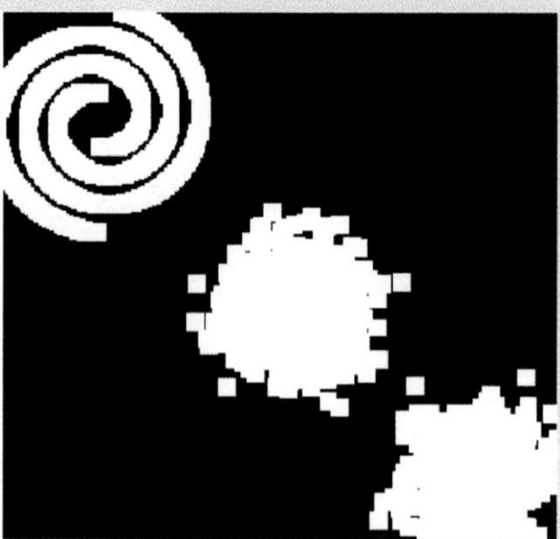

```
#Cross 3x3
imgDilated = cv.dilate(imgBin, cv.getStructuringElement(cv.MORPH_CROSS, (3,3)))
```

```
cv_imshow(imgDilated)
```

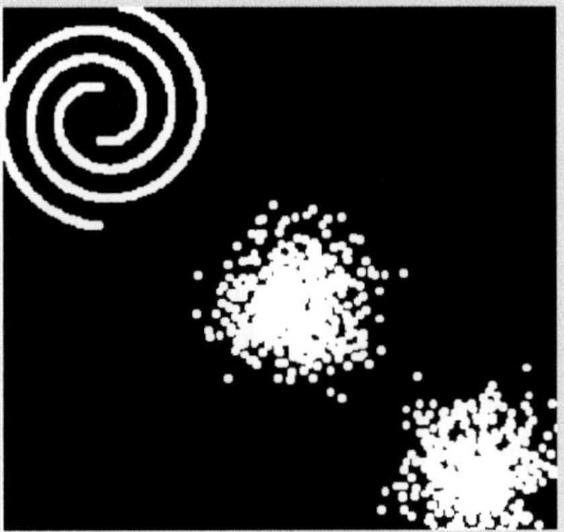

```
#Cross 9x9
imgDilated = cv.dilate(imgBin, cv.getStructuringElement(cv.MORPH_CROSS, (9,9)))
cv_imshow(imgDilated)
```

```
# Customized structural element (kernel):
```

```
#kernel must be of type uint8
```

```
#kernel line type

kernel = np.array([[0, 0, 0, 0, 1],
             [0, 0, 0, 1, 0],
             [0, 0, 1, 0, 0],
             [0, 1, 0, 0, 0],
             [1, 0, 0, 0, 0]]).astype(np.uint8)
imgDilated = cv.dilate(imgBin,kernel)
cv_imshow(imgDilated)
```

```
# Erosion

#Ellipse 3x3
imgEroded = cv.erode(imgBin, cv.getStructuringElement(cv.MORPH_ELLIPSE, (3,3)))
cv_imshow(imgEroded)
```

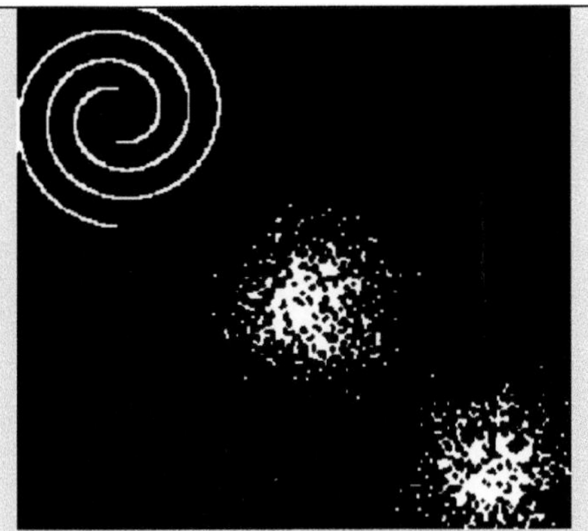

#Ellipse 9x9

```python
imgEroded = cv.erode(imgBin, cv.getStructuringElement(cv.MORPH_ELLIPSE, (9,9)))
cv_imshow(imgEroded)
```

#Rectangle 3x3

```python
imgEroded = cv.erode(imgBin, cv.getStructuringElement(cv.MORPH_RECT, (3,3)))
cv_imshow(imgEroded)
```

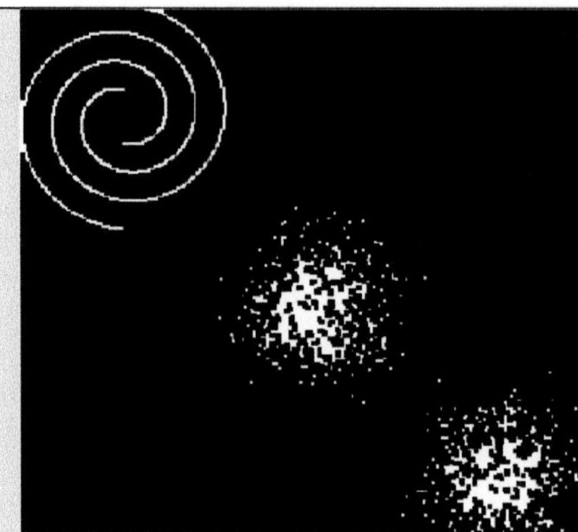

imgEroded = cv.erode(imgBin, cv.getStructuringElement(cv.MORPH_RECT, (9,9)))

cv_imshow(imgEroded)

imgEroded = cv.erode(imgBin, cv.getStructuringElement(cv.MORPH_CROSS, (3,3)))

cv_imshow(imgEroded)

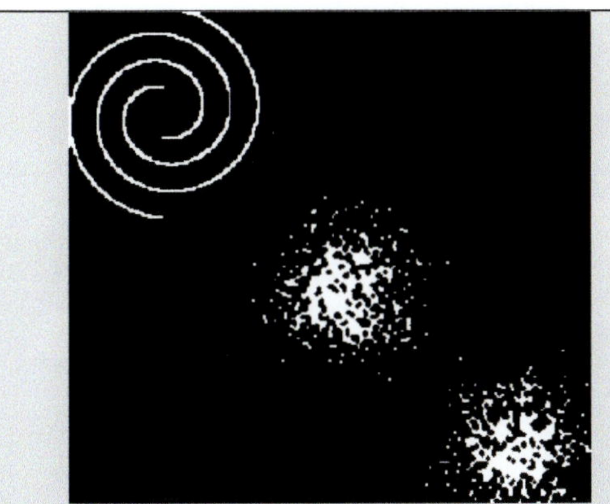

#Cross 9x9

imgEroded = cv.erode(imgBin, cv.getStructuringElement(cv.MORPH_CROSS, (9,9)))

cv_imshow(imgEroded)

Morphological Gradient

img = cv.imread('figuras.png',cv.IMREAD_GRAYSCALE)

img = cv.resize(img,dsize=(0,0),fx=0.3,fy=0.3)

th,imgBin = cv.threshold(img,thresh=250,maxval=255,type=cv.THRESH_BINARY)

```
imgBin = 255 - imgBin
cv_imshow(imgBin)
```

```
imgDilated = cv.dilate(imgBin, cv.getStructuringElement(cv.MORPH_ELLIPSE, (3,3)))
imgEroded = cv.erode(imgBin, cv.getStructuringElement(cv.MORPH_ELLIPSE, (3,3)))
imgGradient = imgDilated - imgEroded
cv_imshow(imgGradient)
```

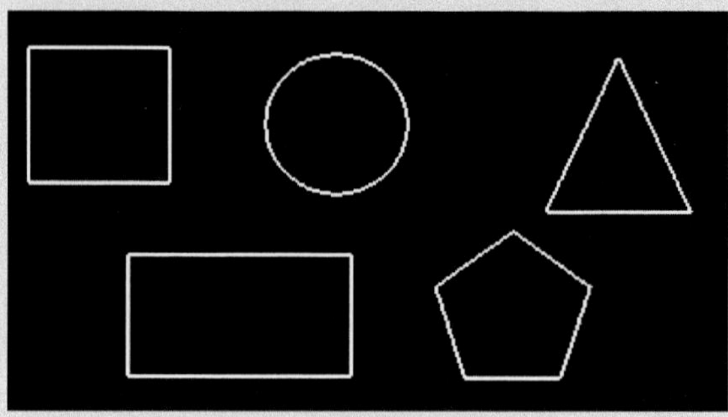

```
# Another way to do this is with the cv.morphologyEx function.
imgGradient = cv.morphologyEx(imgBin, op=cv.MORPH_GRADIENT,
kernel=cv.getStructuringElement(cv.MORPH_ELLIPSE,(3,3)))
cv_imshow(imgGradient)
```

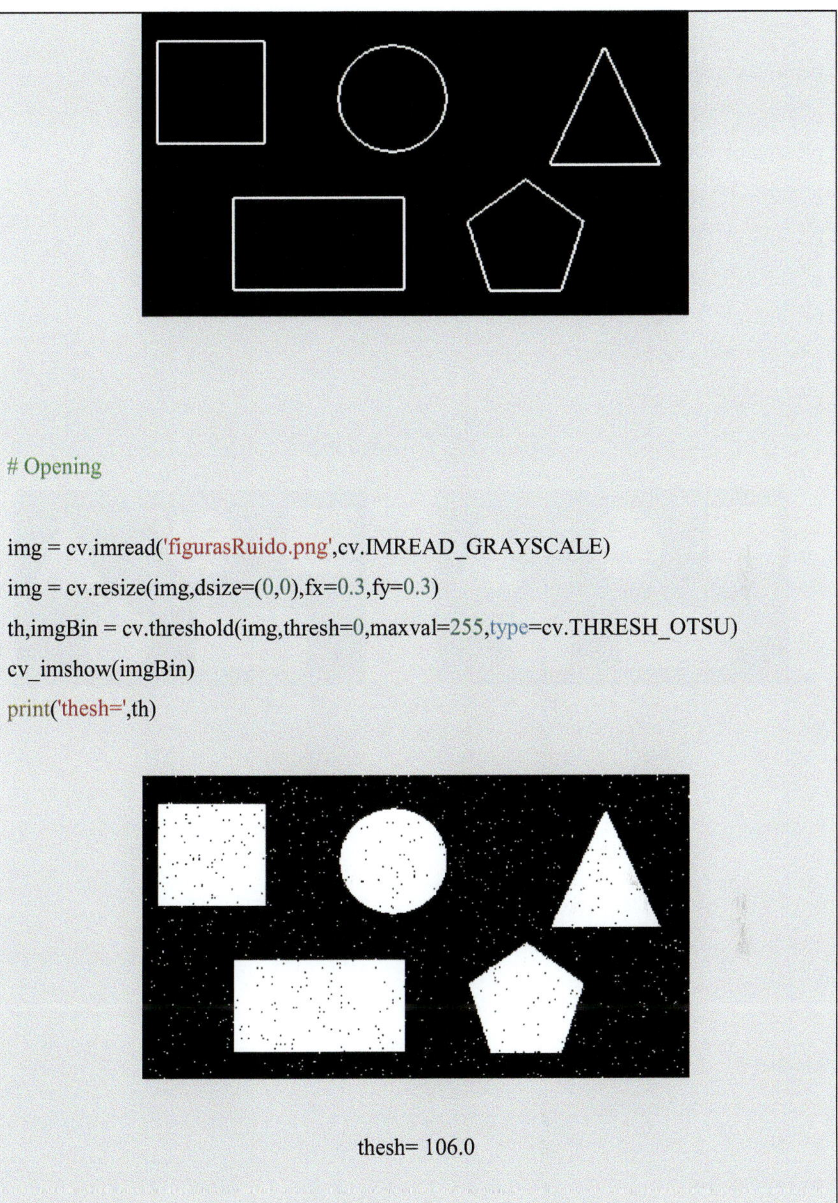

```python
# Opening

img = cv.imread('figurasRuido.png',cv.IMREAD_GRAYSCALE)
img = cv.resize(img,dsize=(0,0),fx=0.3,fy=0.3)
th,imgBin = cv.threshold(img,thresh=0,maxval=255,type=cv.THRESH_OTSU)
cv_imshow(imgBin)
print('thesh=',th)
```

thesh= 106.0

```python
imgOpen = cv.morphologyEx(imgBin, op=cv.MORPH_OPEN,
kernel=cv.getStructuringElement(cv.MORPH_ELLIPSE,(5,5)))
plt.figure(figsize=(10,10))
plt.subplot(1,2,1)
plt.title('Original')
plt.imshow(imgBin,cmap='gray')
plt.subplot(1,2,2)
plt.title('Opening')
plt.imshow(imgOpen,cmap='gray')
plt.show()
```

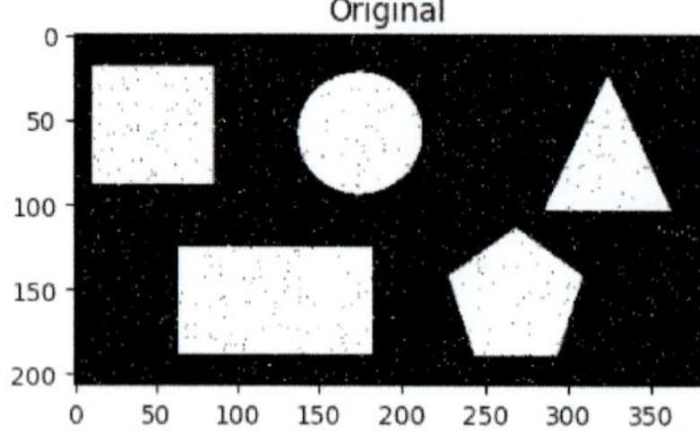
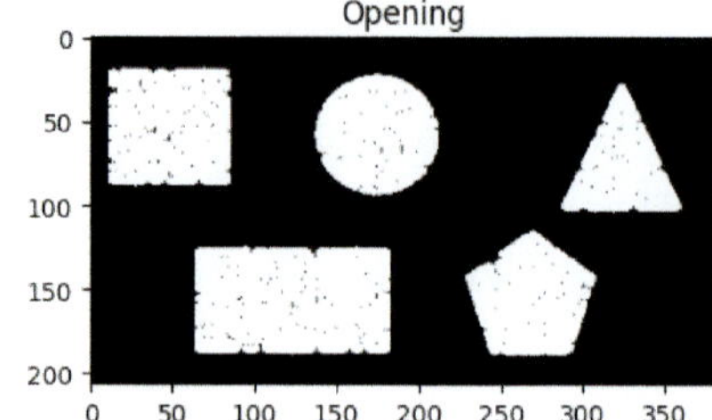

```python
# Closing

imgClose = cv.morphologyEx(imgBin, op=cv.MORPH_CLOSE,
kernel=cv.getStructuringElement(cv.MORPH_ELLIPSE,(5,5)))
plt.figure(figsize=(10,10))
plt.subplot(1,2,1)
plt.title('Original')
plt.imshow(imgBin,cmap='gray')
plt.subplot(1,2,2)
plt.title('Closing')
plt.imshow(imgClose,cmap='gray')
```

```
plt.show()
```

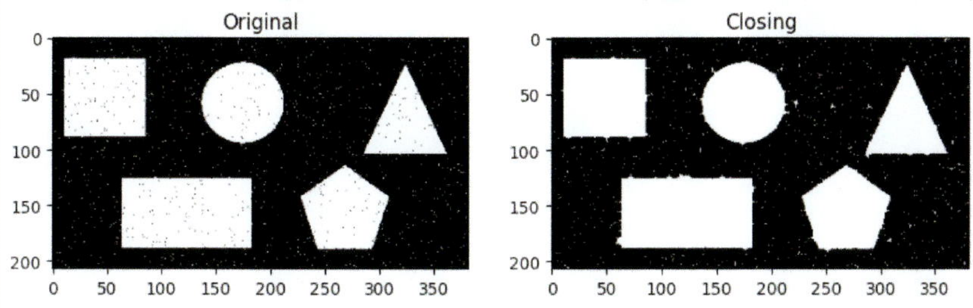

```
# Noise elimination application

# To eliminate noise, an opening operation is applied, followed by a closing operation, or
vice versa.

imgDenoised = cv.morphologyEx(imgOpen, op=cv.MORPH_CLOSE,
kernel=cv.getStructuringElement(cv.MORPH_ELLIPSE,(5,5)))
plt.figure(figsize=(10,10))
plt.subplot(1,2,1)
plt.title('Original')
plt.imshow(imgBin,cmap='gray')
plt.subplot(1,2,2)
plt.title('Denoised')
plt.imshow(imgDenoised,cmap='gray')
plt.show()
```

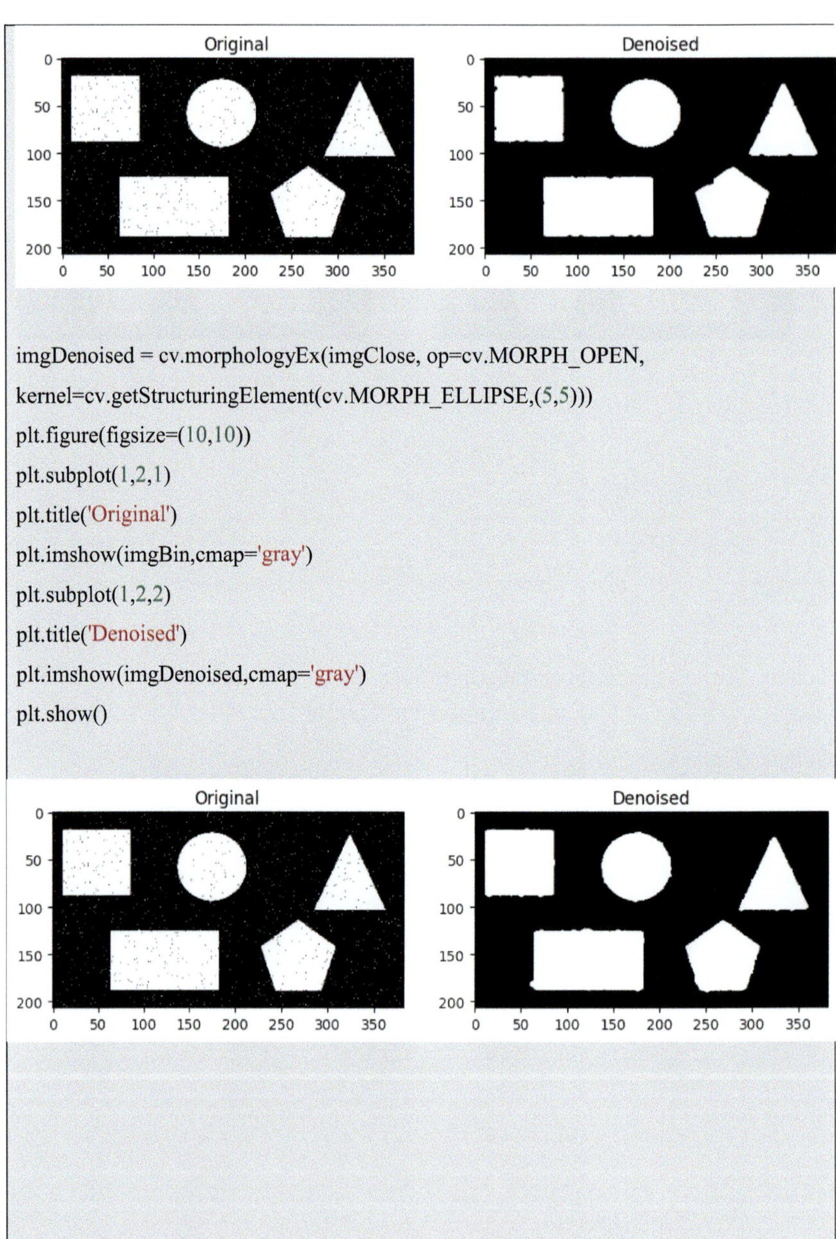

```python
imgDenoised = cv.morphologyEx(imgClose, op=cv.MORPH_OPEN,
kernel=cv.getStructuringElement(cv.MORPH_ELLIPSE,(5,5)))
plt.figure(figsize=(10,10))
plt.subplot(1,2,1)
plt.title('Original')
plt.imshow(imgBin,cmap='gray')
plt.subplot(1,2,2)
plt.title('Denoised')
plt.imshow(imgDenoised,cmap='gray')
plt.show()
```

References

1. Serra, J. (1982). *Image analysis and mathematical morphology*. Academic Press.
2. Soille, P. (2003). Morphological image analysis: principles and applications (2nd ed.). Springer.
3. Gonzalez, R. C., & Woods, R. E. (2018). Digital image processing (4th ed.). Pearson.

4. Haralick, R. M., Sternberg, S. R., & Zhuang, X. (1987). Image analysis using mathematical morphology. *IEEE Transactions on Pattern Analysis and Machine Intelligence, 9*(4), 532–550.
5. Sonka, M., Hlavac, V., & Boyle, R. (2014). Image processing, analysis, and machine vision (4th ed.).